my **revision** notes

OCR AS/A-level Year 1

CHEMISTRY A

Mike Smith

HODDER
EDUCATION
AN HACHETTE UK COMPANY

Hachette UK's policy is to use papers that are natural, renewable and recyclable products and made from wood grown in sustainable forests. The logging and manufacturing processes are expected to conform to the environmental regulations of the country of origin.

Orders: please contact Bookpoint Ltd, 130 Milton Park, Abingdon, Oxon OX14 4SB. Telephone: (44) 01235 827720. Fax: (44) 01235 400454. Email education@bookpoint.co.uk

Lines are open from 9 a.m. to 5 p.m., Monday to Saturday, with a 24-hour message answering service. You can also order through our website: www.hoddereducation.co.uk

ISBN: 978 1 4718 4210 8

© Mike Smith 2015

First published in 2015 by

Hodder Education,
An Hachette UK Company
Carmelite House
50 Victoria Embankment
London EC4Y 0DZ
www.hoddereducation.co.uk

Impression number   10 9 8 7 6 5 4 3 2 1

Year   2019 2018 2017 2016 2015

Cover photo reproduced by permission of Sean Gladwell/Fotolia

Typeset in Integra Software Services Pvt. Ltd., Pondicherry, India

Printed in Spain

A catalogue record for this title is available from the British Library.

Get the most from this book

Everyone has to decide his or her own revision strategy, but it is essential to review your work, learn it and test your understanding. These Revision Notes will help you to do that in a planned way, topic by topic. Use this book as the cornerstone of your revision and don't hesitate to write in it — personalise your notes and check your progress by ticking off each section as you revise.

Tick to track your progress

Use the revision planner on pages 4 and 5 to plan your revision, topic by topic. Tick each box when you have:
- revised and understood a topic
- tested yourself
- practised the exam questions and gone online to check your answers and complete the quick quizzes

You can also keep track of your revision by ticking off each topic heading in the book. You may find it helpful to add your own notes as you work through each topic.

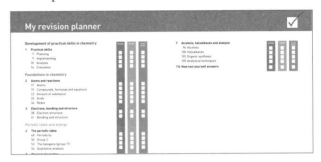

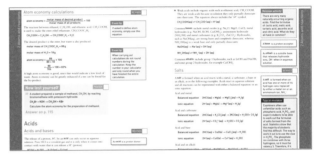

Features to help you succeed

Exam tips

Expert tips are given throughout the book to help you polish your exam technique in order to maximise your chances in the exam.

Typical mistakes

The author identifies the typical mistakes candidates make and explain how you can avoid them.

Now test yourself

These short, knowledge-based questions provide the first step in testing your learning. Answers are at the back of the book.

Definitions and key words

Clear, concise definitions of essential key terms are provided where they first appear.

Key words from the specification are highlighted in bold throughout the book.

Revision activities

These activities will help you to understand each topic in an interactive way.

Exam practice

Practice exam questions are provided for each topic. Use them to consolidate your revision and practise your exam skills.

Summaries

The summaries provide a quick-check bullet list for each topic.

Online

Go online to check your answers to the exam questions and try out the extra quick quizzes at **www.hoddereducation.co.uk/myrevisionnotes**

My revision planner

REVISED   TESTED   EXAM READY

Exam practice answers and quick quizzes at **www.hoddereducation.co.uk/myrevisionnotes**

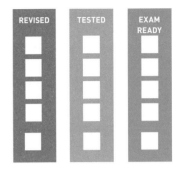

Countdown to my exams

6–8 weeks to go

- Start by looking at the specification — make sure you know exactly what material you need to revise and the style of the examination. Use the revision planner on pages 4 and 5 to familiarise yourself with the topics.
- Organise your notes, making sure you have covered everything on the specification. The revision planner will help you to group your notes into topics.
- Work out a realistic revision plan that will allow you time for relaxation. Set aside days and times for all the subjects that you need to study, and stick to your timetable.
- Set yourself sensible targets. Break your revision down into focused sessions of around 40 minutes, divided by breaks. These Revision Notes organise the basic facts into short, memorable sections to make revising easier.

REVISED ☐

2–5 weeks to go

- Read through the relevant sections of this book and refer to the exam tips, exam summaries, typical mistakes and key terms. Tick off the topics as you feel confident about them. Highlight those topics you find difficult and look at them again in detail.
- Test your understanding of each topic by working through the 'Now test yourself' questions in the book. Look up the answers at the back of the book.
- Make a note of any problem areas as you revise, and ask your teacher to go over these in class.
- Look at past papers. They are one of the best ways to revise and practise your exam skills. Write or prepare planned answers to the exam practice questions provided in this book. Check your answers online and try out the extra quick quizzes at **www.hoddereducation.co.uk/ myrevisionnotes**
- Use the revision activities to try out different revision methods. For example, you can make notes using mind maps, spider diagrams or flash cards.
- Track your progress using the revision planner and give yourself a reward when you have achieved your target.

REVISED ☐

1 week to go

- Try to fit in at least one more timed practice of an entire past paper and seek feedback from your teacher, comparing your work closely with the mark scheme.
- Check the revision planner to make sure you haven't missed out any topics. Brush up on any areas of difficulty by talking them over with a friend or getting help from your teacher.
- Attend any revision classes put on by your teacher. Remember, he or she is an expert at preparing people for examinations.

REVISED ☐

The day before the examination

- Flick through these Revision Notes for useful reminders, for example the exam tips, exam summaries, typical mistakes and key terms.
- Check the time and place of your examination.
- Make sure you have everything you need — extra pens and pencils, tissues, a watch, bottled water, sweets.
- Allow some time to relax and have an early night to ensure you are fresh and alert for the examinations.

REVISED ☐

My exams

AS Chemistry

Date:...

Time:...

Location:...

A-level Chemistry

Date:...

Time:...

Location:...

1 Practical skills

Chemistry is a practical subject so the development of practical skills is essential. These skills will be assessed by your teachers, not under exam conditions. During the course you will be required to carry out a number of experiments, a minimum of 12, which are separately assessed. The experiments that you carry out will cover a range of technical skills and practical apparatus. Teachers will award a pass (or fail) to their students, and performance in this component will be integral to, and examined in, all components of the one-year and the full two-year course.

The assessment will cover four key areas: planning, implementing, analysis and evaluation.

Table 1.1 gives a brief outline of the sorts of experiments you might be expected to encounter.

Table 1.1

Year 1	Year 2
Mole determination	Rates of reaction
Acid–base titration	pH measurement
Qualitative testing of ions	Electrochemical cells
Preparation of an organic liquid	Redox titrations
	A range of organic syntheses and organic analysis
You will be expected to develop your research skills throughout the course.	

Planning

Planning a scientific investigation is an essential skill and enables priorities to be dealt with in a controlled manner instead of simply reacting to things as they come along.

It is essential that you are able to identify the key stages in a scientific investigation and select the appropriate:
● reagents and conditions
● practical technique(s) and apparatus

When you have decided on the chemicals and apparatus that you will need you must carry out a **risk assessment** to ensure that appropriate precautions are put in place to allow the chemicals to be handled safely. Details of individual hazards can be obtained from **www.cleapss.org. uk/secondary/secondary-science/hazcards**. Whenever carrying out a chemical experiment it is usual to wear a laboratory coat and possibly protective gloves, but it is essential to *wear safety glasses at all times* when handling chemicals.

Now test yourself

TESTED

1 Figure 1.1 shows a number of hazard warning labels. Use the internet or look up a chemical catalogue to help you identify which hazard is represented by each warning label.

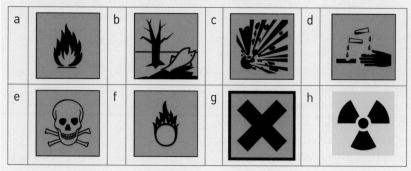

Figure 1.1

Note: Some chemicals are too hazardous to be used in schools, therefore not all of these warning labels will be found in your chemistry laboratory.

Answer on p. 114

Revision activity

Use either your schools database (probably CLEAPSS) or the internet to decide on what safety precautions you would need to take if you were carrying out an experiment using:
(a) sodium hydroxide
(b) hydrogen peroxide
(c) ethanoyl chloride

Identifying practical techniques and equipment

REVISED

A crucial stage in planning a scientific investigation is to identify the most appropriate practical technique to allow you to safely carry out your experiment.

Now test yourself

TESTED

2 Match up the **number** of each laboratory technique in Table 1.2 to the **letter** of its most appropriate use.

Table 1.2

	Technique		Use
1	Filtration	A	Weighing out chemicals
2	Water bath	B	Determining the concentration of a solution
3	Distillation	C	Heating an aqueous solution rapidly to 60°C
4	Balance	D	Separating a mixture of liquids
5	Bunsen burner	E	Collecting a gas that is insoluble in water
6	Gas syringe	F	Separating a solid from a liquid
7	Titration	G	Heating a flammable liquid to 120°C
8	Collecting a gas over water	H	Maintaining a reaction at 50°C
9	Heating mantle	I	Collecting a water-soluble gas

Answer on p. 114

Having identified the most appropriate technique for a particular scientific investigation, apparatus must then be selected to allow that technique to be carried out effectively and safely.

Implementing

For most practicals that you carry out you will be provided with a set of instructions and it is essential that you follow these.

Chemical experiments are a bit like recipes, and for them to work you have to follow the recipe precisely. However, each piece of laboratory apparatus used to measure a quantity has a limit to its precision. For example, a fairly standard balance may give a measurement to 2 decimal places but as you are using it the second decimal place often fluctuates and may change. This indicates that the balance has an inbuilt error.

A useful rule of thumb for any apparatus is that the error range will be +/− half of the smallest digit. Table 1.3 shows an example.

Table 1.3

Balance	Mass	Error range	Range
1 decimal place	2.4 g	+/− 0.05	2.35–2.45 g
2 decimal places	2.40 g	+/− 0.005	2.395–2.405 g
3 decimal places	2.400 g	+/− 0.0005	2.3995–2.4005 g

For most measuring equipment, the manufacturer will give the maximum error that is inherent in using that piece of apparatus; this is sometimes etched onto the apparatus but, in other cases, will need to be looked up.

The **percentage error** for each balance is shown below:

1-decimal-point balance $\qquad \dfrac{0.05}{2.4} \times 100 = 2.08\%$

2-decimal-point-balance $\qquad \dfrac{0.005}{2.40} \times 100 = 0.208\%$

3-decimal-point balance $\qquad \dfrac{0.0005}{2.400} \times 100 = 0.0208\%$

The more accurate the apparatus, the lower the percentage error.

The maximum error is an inevitable part of using that piece of equipment and is distinct from the competence with which the experiment is carried out. Measuring a volume in an apparatus, such as a pipette, only requires one reading, so only one error is incurred (the error range, usually +/− 0.05 cm³, is etched onto the pipette). However, measuring a volume in a burette requires two readings — the initial volume and the final volume, so the maximum error is doubled.

Important decisions are sometimes based on the results of experiments. For example, titrations are used in health care, in the food industry and in forensic science. It is crucial that the people making decisions based on the results obtained understand the extent to which they can rely on the data from their analysis.

> **Percentage error =**
>
> $\dfrac{\text{maximum error}}{\text{actual value}} \times 100$

> **Revision activity**
>
> You have to measure 24 cm³ of liquid using either a 10 cm³ or a 25 cm³ measuring cylinder. Check the glassware in your laboratory to find the error range and decide which would be the most accurate.

Volumetric equipment

REVISED

It is important to stress that when using volumetric apparatus the correct position of the meniscus is essential. In Figure 1.2, (a) is the correct way to measure volume, taking account of the meniscus. All volumetric apparatus (pipettes, burettes and volumetric flasks) are manufactured such that the correct volume is obtained when the bottom of the meniscus sits exactly on the line.

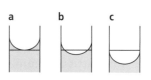

Figure 1.2

Now test yourself

3 (a) Calculate the percentage error in measuring $50\,cm^3$ of a solution using:
  (i) a $250\,cm^3$ measuring cylinder with a maximum error of $1\,cm^3$
  (ii) a $100\,cm^3$ measuring cylinder with a maximum error of $0.5\,cm^3$
  (iii) a $50\,cm^3$ pipette with a maximum error of $0.1\,cm^3$
  (iv) a $25\,cm^3$ pipette with a maximum error of $0.06\,cm^3$, used twice
  (v) a $50\,cm^3$ burette, where each reading has a maximum error of $0.05\,cm^3$
 (b) Calculate the percentage error in measuring a temperature rise of 6°C using a thermometer with a maximum error of:
  (i) 0.5°C
  (ii) 0.1°C

Answer on p. 114

Recording results

When recording data, the precision should be indicated appropriately. For example, if you use a balance that reads to 2 decimal places, the masses recorded should indicate this. This may seem obvious for a mass of, for example, 24.79 g. However, you must remember that this applies equally for a mass of, for example, 24.80 g. Here the '0' should be included after the '8' to indicate that this mass is also precise to 2 decimal places. Recording the mass as 24.8 g is incorrect and will be penalised in an exam.

Burette readings are normally recorded to $0.05\,cm^3$ as this represents the appropriate maximum error. In Figure 1.3 a reading of $12.60\,cm^3$ or of $12.65\,cm^3$ is acceptable, but $12.64\,cm^3$ or $12.6\,cm^3$ are not.

The maximum error can be regarded as $+/-$ one half of the smallest division, which for a standard burette is $0.10\,cm^3$, so the error is $+/- 0.05\,cm^3$. The volume measured in a burette should always be recorded to 2 decimal places and the second decimal place must always be either '0' or '5'.

Analysis

When carrying out experiments you will be expected to interpret both qualitative and quantitative data.

In the first year of the course the qualitative analysis that you are expected to know is listed in section 3.1.4 of the specification, which details a range of chemical tests for ions including: carbonate (CO_3^{2-}), sulfate (SO_4^{2-}), halides (Cl^-, Br^-, I^-) and ammonium (NH_4^+).

You will encounter quantitative analysis when carrying out experiments involving moles (see pages 22–27) or enthalpy changes, which occur later in the course.

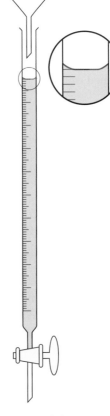

Figure 1.3

Significant figures

In some cases the number of significant figures is simply the number of digits in the answer. 72.67 has 4 significant figures, while 72.7 has 3 significant figures and 73 has just 2.

In other cases numbers may need rounding up or down before quoting the answer to a particular number of significant figures. 94.64 has 4 significant figures but to 3 significant figures this is 94.6 (as 94.64 is nearer to 94.6 than to 94.7). A number ending in a '5' is rather arbitrarily raised to the number above. So 11.5 must be written as 12 when quoted to 2 significant figures.

When a number such as 0.00461 has '0's *after* the decimal point these are not considered as 'significant'. 0.00461 therefore has 3 significant figures.

When a number has '0's *before* the decimal point they are significant, such that 7630.00 has 4 significant figures (Figure 1.4).

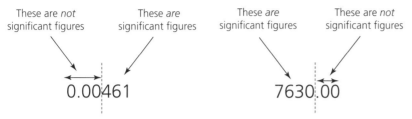

Figure 1.4

A number such as 1950 has 4 significant figures but if you are asked to quote this to 2 significant figures there is a temptation to quote this as '2000', which is incorrect as 2000 has 4 significant figures. The way round it is to write the number in standard (index) form i.e. 2.0×10^3. 1950 written in standard form would be 1.95×10^3, which is to 3 significant figures.

Often your calculator will display an answer containing more digits than you were given in the data.

Suppose you were asked to calculate the concentration of HCl(aq) when $24.2\,cm^3$ of the HCl(aq) was neutralised by $25.0\,cm^3$ of $0.500\,mol\,dm^{-3}$ NaOH(aq). If you did this calculation correctly your calculator would show the concentration to be $0.516528925\,mol\,dm^{-3}$. The concentration of the solution is not known to this degree of precision.

The accuracy should be limited to the precision of the data or, in an experiment, the accuracy of the apparatus. In the example above the data are given to 3 significant figures and so the answer should also be limited to 3 significant figures. The figures after the third are dropped and the number is *rounded*. 0.516528925 when rounded to 3 significant figures is 0.517.

When carrying out a calculation always quote your answer to the same number of significant figures as given in the data. If the number of significant figures in the data varies, the least accurate should be used.

Exam tip

When carrying out calculations it is essential that you do not round until the end of the calculation. If necessary use the 'memory' function on your calculator.

Using numbers in standard (index) form

REVISED

Numbers can be written in different formats. A common way to write numbers is to use the decimal notation, for example 123642.78 and 0.0005432. When working with very large numbers (123642.78) or very small numbers (0.0005432) it is convenient to write these in **standard notation**. This means writing the number as a product of two factors:
- in the first factor the decimal point *always* comes after the first digit
- the second factor is *always* a multiple of 10

Example

Number	9874	987.4	98.74	9.874	0.9874	0.09874	0.009874
Standard form	9.874×10^3	9.874×10^2	9.874×10^1	$9.874 \times 10^{0*}$	9.874×10^{-1}	9.874×10^{-2}	9.874×10^{-3}

* Since $10^0 = 1$, it follows that 9.874×10^0 is normally written as 9.874.

Now test yourself

4 (a) Write the following numbers to 3 significant figures.
   (i) 734.8
   (ii) 698.456
   (iii) 0.0003456
  (b) Write the following numbers to 2 significant figures and in standard form.
   (i) 734.8
   (ii) 698.456
   (iii) 0.0003456

Answer on p. 114

Answer on p. 114

Exam tip

It is always useful to estimate the answer to a calculation before doing the calculation on a calculator. It makes it easy to spot whether or not you have input the data into the calculator correctly.

Drawing graphs

REVISED

Figure 1.5 shows a simple relationship between two variables.

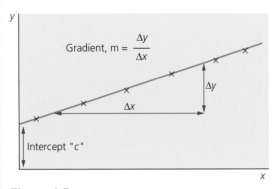

Figure 1.5

The relationship between x and y is $y = mx + c$ (where m is the gradient and c is the intercept). Δx is the change in x and Δy is the change in y. The gradient, m, can be calculated by using $m = \Delta y / \Delta x$.

When drawing graphs you should:
1 Choose a scale that will allow the graph to cover as much of the graph paper as possible. It is helpful to start both axes at zero but if all the points on one axis are between 90 and 100, to start at zero on that axis would cramp your graph into a small section of the paper (Figure 1.6a). In this case it is much better to truncate the x-axis so that the graph fills as much of the paper as possible (Figure 1.6b).

Exam tip

The symbol Δ is used to represent 'change in…', such that ΔT is change in temperature, ΔP is change in pressure and ΔV is change in volume.

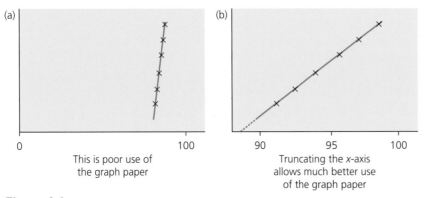

Figure 1.6

2 Label the axes with the dimensions and the units such as:
- ○ Volume/cm³
- ○ Concentration/mol dm⁻³

3 After plotting all the points on a graph often you may not get a perfect straight line or a curve that goes through all of the points. You have to draw a line of best fit for the points (Figure 1.7).

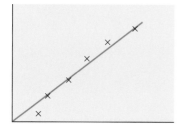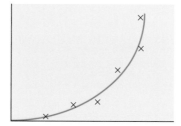

Figure 1.7

4 Figure 1.8 shows how to draw tangents to a curve.

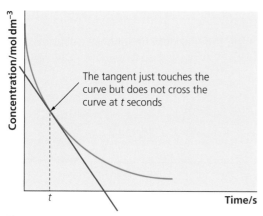

Figure 1.8

By calculating the gradient of the tangent it is possible to work out how the concentration changes with respect to time after t seconds. This enables you to calculate the rate of reaction after t seconds. The units of rate are the units of y/x, which are mol dm⁻³/s, which is written as mol dm⁻³ s⁻¹.

Revision activity

A student reacted 0.42 g sodium hydrogen carbonate, $NaHCO_3$(s) with excess dilute HCl(aq) and measured the volume of CO_2(g) evolved every minute.

$$NaHCO_3(s) + HCl(aq) \rightarrow NaCl(aq) + H_2O(l) + CO_2(g)$$

The results are recorded in Table 1.4.

Table 1.4

Time/min	0	1	2	3	4	5	6	7	8
Volume of CO_2(g)/cm³	0	40	71	96	105	114	118	120	120

Plot the results, label the axes, and deduce the rate of reaction after:
(a) 2 minutes
(b) 4 minutes

Evaluation

You should be able use your knowledge and understanding to evaluate your results and use them to draw valid conclusions. You should be able to identify anomalies in experiment data.

TESTED

Now test yourself

5 A student carries out a titration and records his results, as shown in Table 1.5.

Table 1.5

Titration	Volume/cm³
Rough	23
1	23.50
2	24.50
3	23.60

He wants to work out the average titre value. Which results in the table should he ignore? Calculate the average titre.

Answer on p. 114

Limitations in experimental procedure

REVISED

Apart from the limitations imposed by apparatus, experiments also have errors caused by the procedure adopted. Such errors are difficult to quantify, but the following check list might help you to assess an experiment.

Purity of chemicals

Can you be sure that the substances you are using are pure? Solids may be damp and if a damp solid is weighed, the absorbed moisture is included in the mass.

Does the experiment involve a reactant or a product that could react with the air? Remember that air contains carbon dioxide, which is acidic and reacts with alkalis. Some substances react with the oxygen present. Air is also always damp.

Heating substances

If you need to heat a substance until it decomposes, you can only be sure that the decomposition is complete if you heat to constant mass.

Is it possible that heating is too strong and the product has decomposed further?

Solutions

If an experiment is quantitative, can you guarantee that any solution used is exactly at the stated concentration?

Gases

If a gas is collected during the experiment, can you be sure that none has escaped? If you collect a gas over water, are you sure it is not soluble?

If you collect the gas in a gas syringe the volume of the gas is temperature-dependent.

Timing

If an experiment involves timing, are you sure that you can start and stop the timing exactly when required?

Enthalpy experiments

Heat loss is always a problem in enthalpy experiments, particularly if the reaction is slow.

Improving experimental design

REVISED

It is useful to look at all measurements and to calculate the percentage error in each. Is there any measurement whose percentage error is significantly larger than the others? If there is it is worth considering using more precise apparatus for that measurement.

It is a mistake to imagine that perfection can be achieved by using more complicated apparatus if the fault lies in the method that is being employed.

If the problem is that a gas to be collected over water is slightly soluble then a change in the method is appropriate. Using a gas syringe would be an effective remedy.

If the problem with an enthalpy reaction is that it is too slow, maybe using powders or a more concentrated solution would help.

Exam practice

1 Two students were each provided with a small lump of impure magnesium carbonate and each student was asked to design an experiment to determine the percentage purity.
   (a) One student decided to react the impure magnesium carbonate with hydrochloric acid and collect the carbon dioxide evolved by displacement of water.

   $$MgCO_3(s) + 2HCl \rightarrow MgCl_2(aq) + H_2O(l) + CO_2(g)$$

   (i)   Sketch the apparatus that the student would use. [2]
   (ii) Explain how the student would calculate the percentage purity. [4]
   (iii) Identify a procedural error in the method adopted. [1]
   (iv) Suggest an improvement. [1]
   (b) The second student decided to decompose the magnesium carbonate and to weigh the magnesium oxide using a 2-decimal-point balance.

   $$MgCO_3(s) \rightarrow MgO(s) + CO_2(g)$$

   (i) Write a brief method for this experiment. [5]
   (ii) If the final mass of MgO(s) was 0.20 g, calculate the percentage error in weighing this sample. [2]

2 Lithium reacts with water to produce a solution of lithium hydroxide and hydrogen gas.

$2Li(s) + 2H_2O(l) \rightarrow 2LiOH(aq) + H_2(g)$

The apparatus shown in Figure 1.9 was used to monitor the rate of reaction.

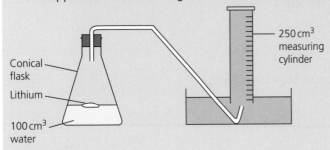

Figure 1.9

A small piece of lithium was weighed on a two decimal point balance and added to the conical flask and the bung quickly replaced. The volume of $H_2(g)$ collected was recorded every 30 seconds. The results are shown in Table 1.6.

Table 1.6

	Error range
Mass of weighing boat = 12.56 g	+/− 0.005 g
Mass of weighing boat + lithium = 12.68 g	+/− 0.005 g

Time/s	30	60	90	120	150	180	210	240
Volume of $H_2(g)$/cm³	60	110	150	180	195	203	208	208

The error range in when reading the volume of the $H_2(g)$ is +/− 1.00 cm³

(a) Plot the results. Draw a tangent at 60 seconds and deduce the rate of reaction after 60 seconds by finding the gradient of the tangent at 60 seconds. [4]

(b) Calculate the percentage error in:
  (i) the mass of lithium
  (ii) the volume of hydrogen after 240 seconds [4]

(c) Suggest an improvement to the method. Justify your suggestion. [3]

Answers and quick quiz 1 online

ONLINE

Summary

You should now have an understanding of:
- planning — experimental design and the need to ensure that safe practices are adopted
- implementing — how to use practical apparatus and techniques correctly
- recording — observations and results recorded appropriately

- analysis — interpreting experimental qualitative and quantitative data
- evaluation — how to draw conclusions, estimate errors and suggest valid improvements

2 Atoms and reactions

Atoms

The protons, neutrons and electrons that make up atoms are described in Table 2.1.

Table 2.1

Particle	Relative mass	Relative charge	Distribution
Proton, p	1	1+	Nucleus
Neutron, n	1	0	Nucleus
Electron, e	1/1836	1–	Orbits/shells

The **atomic number** and the **mass number** can be used to deduce the number of protons, neutrons and electrons in atoms and in ions.

Atoms are neutral and contain the same number of protons as electrons. Positive ions have lost electrons and hence have more protons than electrons; negative ions have gained electrons so they have fewer protons than electrons (Figure 2.1).

> **Atomic number** is the number of protons in an atom or an ion of an element.
>
> **Mass number** is the number of protons plus neutrons in the nucleus of an atom.
>
> (Mass number minus atomic number) tells us the number of neutrons in an atom.

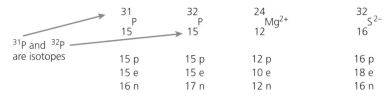

$^{31}_{15}P$	$^{32}_{15}P$	$^{24}_{12}Mg^{2+}$	$^{32}_{16}S^{2-}$
^{31}P and ^{32}P are isotopes			
15 p	15 p	12 p	16 p
15 e	15 e	10 e	18 e
16 n	17 n	12 n	16 n

Figure 2.1

Chlorine has two **isotopes**: ^{35}Cl and ^{37}Cl.

> **Isotopes** are atoms of the same element with different masses — they have the same number of protons (and electrons) but different number of neutrons.

Relative masses

Most exam papers ask for at least one or two definitions, which might include definitions of **relative isotopic mass** or **relative atomic mass**.

It is possible to use the definition of relative atomic mass and amend it slightly to create definitions for:

- **relative molecular mass** which applies to all covalent molecules
- **relative formula mass** which applies to all compounds

> **Relative molecular mass** is the weighted mean mass of a molecule compared with 1/12th of the mass of a ^{12}C atom.
>
> **Relative formula mass** is the weighted mean mass of a formula unit compared with 1/12th of the mass of a ^{12}C atom.

> The **relative isotopic mass** is the mass of an atom/isotope of the element compared with 1/12th of the mass of a ^{12}C atom whose mass is exactly 12.
>
> The **relative atomic mass** of an element is the weighted mean mass of an atom of the element compared with 1/12th of the mass of a ^{12}C atom whose mass is exactly 12.

Calculations

Relative atomic mass

Example 1

A sample of iron contains three isotopes: ^{54}Fe, ^{56}Fe and ^{57}Fe. The relative abundance is 2 : 42 : 1 respectively. Calculate the relative atomic mass.

Answer

Three isotopes, therefore three brackets, each containing the mass number of the isotope, multiplied by the relative amount of each

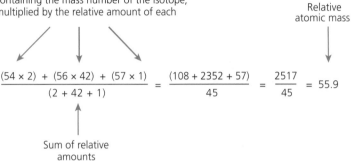

$$\frac{(54 \times 2) + (56 \times 42) + (57 \times 1)}{(2 + 42 + 1)} = \frac{(108 + 2352 + 57)}{45} = \frac{2517}{45} = 55.9$$

Sum of relative amounts

Relative atomic mass

Example 2

Ne has two isotopes, ^{20}Ne and ^{22}Ne, and the relative atomic mass is 20.18. Calculate the percentage of each isotope.

Answer

If the percentage of ^{20}Ne is x% then there must be $(100 - x)$% of ^{22}Ne, such that:

$$\frac{20x + 22(100 - x)}{100} = 20.18$$

$$20x + 2200 - 22x = 2018$$

$$20x - 22x = 2018 - 2200$$

$$-2x = -182$$

$$x = 91\%$$

The sample contains 91% ^{20}Ne and 9% ^{22}Ne.

Relative molecular mass and relative formula mass

Relative molecular mass applies to covalent molecules only. Relative formula mass applies to all substances.

Example 1

Calculate the relative molecular mass of glucose, $C_6H_{12}O_6$.

Answer

$C_6H_{12}O_6 = (6 \times 12.0) + (12 \times 1.0) + (6 \times 16.0) = 72.0 + 12.0 + 96.0 = 180.0$

Example 2

Calculate the relative formula mass of sodium carbonate, Na_2CO_3.

Answer

$Na_2CO_3 = (23.0 \times 2) + 12.0 + (16.0 \times 3) = 46.0 + 12.0 + 48.0 = 106.0$

Example 3

Calculate the relative formula mass of barium chloride crystals, $BaCl_2.2H_2O$.

Answer

$BaCl_2.2H_2O = 137.3 + (2 \times 35.5) + 2(1.0 + 1.0 + 16.0) = 137.3 + 71.0 + 36.0 = 244.3$

Now test yourself

TESTED

1 Deduce the number of protons, neutrons and electrons present in each of the following:
   (a) ^{16}O
   (b) $^{23}Na^+$
   (c) $^{19}F^-$
2 Rubidium consists of two isotopes: Rb-85 has an abundance of 72.2% and Rb-87 has an abundance of 27.8%. Calculate the weighted mean atomic mass of rubidium.
3 A sample of boron was known to contain two different isotopes, $^{10}_{5}B$ and $^{11}_{5}B$. The relative atomic mass of the sample of boron was 10.8. Calculate the percentage of each isotope in the sample.
4 Calculate the relative formula mass of each of the following:
   (a) magnesium hydroxide, $Mg(OH)_2$
   (b) sodium carbonate crystals, $Na_2SO_4.10H_2O$
5 Lithium has two isotopes, ^6Li and ^7Li.
   (a) Define the key term *isotope*.
   (b) Explain how the two isotopes of lithium differ.
   (c) Use the mass spectrum below to calculate the A_r of lithium. Quote your answer to 2 significant figures.

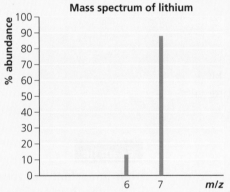

Mass spectrum of lithium

Answers on p. 114

Exam tip

Students often lose marks by careless use of words. For instance, sodium carbonate, Na_2CO_3, must *not* be described as a molecule — it is a giant lattice made up of millions of ions and does not exist as a single Na_2CO_3 molecule. The same applies to $BaCl_2.2H_2O$.

Revision activity

Select an element with atomic number between 21 and 30. Use the internet or a reference book to find the stable isotopes of that element. Calculate its relative atomic mass.

Compounds, formulae and equations

Chemical equations

REVISED

It is essential that you are able to write the formulae of a range of common chemicals and to write balanced equations.

The periodic table can be used to deduce the formula of most chemicals, although there are many exceptions (Table 2.2).

Table 2.2

Group	1	2	13	14	15	16	17
Number of bonds (valency)	1	2	3	4	3	2	1

When a group 1 element forms a compound with a group 16 element:
- the group 1 element (e.g. lithium, Li) forms one bond and the group 16 element (e.g. oxygen, O) forms two bonds. It follows that two lithiums are required for each oxygen. Hence, the formula of the compound is Li_2O.

When a group 13 element forms a compound with a group 16 element:
- the group 13 element (e.g. aluminium, Al) forms three bonds and the group 16 element (e.g. oxygen, O) forms two bonds. It follows that two aluminiums require three oxygens (each supply six bonds). Hence, the formula of the compound is Al_2O_3.

Another way of deducing formulae is to use the 'valency cross-over' technique. Using aluminium oxide as an example, follow these simple steps.

Step 1 Write each of the symbols:

Al O

Step 2 Write the valency at the top right-hand side of the symbol:

Al^3 O^2

Step 3 Cross over the valencies:

$$Al^3_{2} \diagdown O^2_{3}$$

Step 4 Write the crossed-over valencies at the bottom right of the other symbol:

Al_2O_3

This is the formula of the compound.

You are expected to know the formulae of: hydrochloric acid, HCl; sulfuric acid, H_2SO_4; nitric acid, HNO_3; and their corresponding salts.

You should be able to use Table 2.3 to work out most formulae.

Table 2.3 Valencies of elements and groups of elements

1	All group 1 elements, hydrogen (H), silver (Ag) and ammonium (NH_4)	All group 17 elements, hydroxide (OH), nitrate (NO_3), hydrogencarbonate (HCO_3)
2	All group 2 elements, iron (Fe), copper (Cu), zinc (Zn), lead (Pb), tin (Sn)	Oxygen (O), sulfur (S), sulfate, (SO_4) carbonate, (CO_3)
3	All group 13 elements, iron (Fe)	
4	Carbon (C), silicon (Si), lead (Pb), tin, (Sn)	

Exam tip

It is important that you learn all of these valencies — if you get a formula wrong it usually means you will get the equation wrong and also any subsequent calculations. Getting a formula right means you are less likely to lose marks in any calculation that follows.

Writing equations

The importance of being able to provide the correct formulae for substances is that it enables you to write equations for chemical reactions that take place. An equation not only summarises the reactants used and the products obtained but it also indicates the numbers of particles of each substance that are required or produced.

A very simple case is the reaction of carbon and oxygen to make carbon dioxide. This is summarised in an equation as:

$$C + O_2 \rightarrow CO_2$$

Exam practice answers and quick quizzes at www.hoddereducation.co.uk/myrevisionnotes

It tells us that one atom of carbon reacts with one molecule of oxygen to make one molecule of carbon dioxide.

The reaction between carbon and hydrogen is:

$$C + 2H_2 \rightarrow CH_4$$

This means that two molecules of hydrogen are required for each atom of carbon in order to make one molecule of methane, CH_4.

In any equation all symbols must be balanced. You may be asked to include state symbols: (g), (l), (s) or (aq).

Now test yourself

TESTED ☐

6 Write the chemical formula of each of the following compounds:
  (a) magnesium chloride
  (b) aluminium sulfate
7 The formula of rubidium chloride is RbCl. What is the formula of rubidium sulfate?
8 The formula of manganese sulfate is $MnSO_4$. What is formula of manganese bromide?
9 Write an equation, including state symbols, for the reaction between:
  (a) zinc oxide solid and aqueous hydrochloric acid
  (b) methane, CH_4, and oxygen to form carbon dioxide and water

Answers on p. 114

You will also be expected to learn and recall the formula of various ions, including those listed in Table 2.4.

Table 2.4 Common ions (cations and anions)

Positive ions (cations)		Negative ions (anions)	
NH_4^+	Ammonium	NO_3^-	Nitrate
Zn^{2+}	Zinc	CO_3^{2-}	Carbonate
Ag^+	Silver	SO_4^{2-}	Sulfate
Fe^{2+}	Iron(II)	OH^-	Hydroxide
Fe^{3+}	Iron(III)	Cl^-, Br^-, I^-	Halides

All ionic compounds are neutral and the charges of the separate ions have to balance.
● The ammonium ion has a charge of 1+ and the hydroxide ion has a charge of 1− so the formula of ammonium hydroxide is NH_4OH.
● The iron(II) ion has a charge of 2+ and the hydroxide ion has a charge of 1− so the formula of iron(II) hydroxide is $Fe(OH)_2$.
● The iron(III) ion has a charge of 3+ and the hydroxide ion has a charge of 1− so the formula of iron(III) hydroxide is $Fe(OH)_3$.

Now test yourself

TESTED ☐

10 What are the formulae of the following ionic compounds:
  (a) silver sulfate
  (b) aluminium nitrate
  (c) iron(III) nitrate

Answer on p. 114

Amount of substance

The mole

The mass of 1 mol of substance = relative formula mass in grams = **molar mass**. The units of molar mass are $g\,mol^{-1}$.

The amount of substance in moles is given the symbol n.

> **The mole** is defined as the amount of substance that contains as many particles as there are atoms in 12 g of the carbon-12 (^{12}C) isotope and is equal to the **Avogadro constant, N_A** = 6.02 × $10^{23}\,mol^{-1}$.

> **Molar mass** is defined as the mass per mole of a substance and is given the symbol M.

Revision activity

A sweet has the dimensions shown in Figure 2.2.

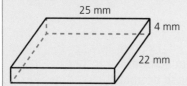

Figure 2.2

(a) Calculate the volume of 1 sweet in (i) mm^3, (ii) m^3, (iii) km^3.
(b) Calculate the volume of 1 mole of the sweets in km^3.
(c) The UK has an area of approximately 243 610 km^2. If 1 mole of the sweets were placed on the UK, calculate the depth of the sweets.

Empirical and molecular formulae

It is important to understand the difference between an **empirical** and a **molecular formula**.

> The **empirical formula** is the simplest whole number ratio of atoms of each element in a compound.

> The **molecular formula** is the actual number of atoms of each element in a molecule of a compound.

Example

Compound **A** has a relative molecular mass of 90 and has a composition by mass of carbon, 26.8%; hydrogen, 2.2%; oxygen, 71.0%. Calculate the empirical formula and the molecular formula.

Divide the % of each element by its own relative atomic mass:

C	:	H	:	O
$\dfrac{26.8}{12.0}$	:	$\dfrac{2.2}{1.0}$	:	$\dfrac{71.0}{16.0}$
2.2	:	2.2	:	4.4

Divide each by the smallest:

$$\frac{2.2}{2.2} = 1 \qquad \frac{2.2}{2.2} = 1 \qquad \frac{4.4}{2.2} = 2$$

Ratio is 1:1:2, hence the empirical formula is $C_1H_1O_2 = CHO_2$.

Deduce how many empirical units are needed to make up the **molecular mass:**

$$CHO_2 = 12.0 + 1.0 + 32.0 = 45.0$$

$$\frac{molecular\ mass}{empirical\ mass} = \frac{90.0}{45.0} = 2.0$$

Therefore, the molecular formula is made up of *two* empirical units. Hence the molecular formula is $C_2H_2O_4$.

When carrying out calculations, students often round numbers in the middle of the calculation instead of at the end. This often leads to an incorrect answer. This is illustrated below in an empirical formula calculation.

A compound contains 40.7% carbon, 5.1% hydrogen and 54.2% oxygen by mass.

Typical mistake						Correct method				
C	:	H	:	O		C	:	H	:	O
$\frac{40.7}{12.0}$	:	$\frac{5.1}{1.0}$	:	$\frac{54.1}{16.0}$		$\frac{40.7}{12.0}$	:	$\frac{5.1}{1.0}$	:	$\frac{54.1}{16.0}$
3.4	:	5.1	:	3.4		3.4	:	5.1	:	3.4
1	:	**1.5**	:	1		1	:	1.5	:	1
1	:	**2**	:	1		2	:	3	:	2

Hence, empirical formula is CH_2O

Hence, empirical formula is $C_2H_3O_2$

The mistake occurs here when 1.5 is rounded up to 2. You should always look to see if a simple multiple can lead to a set of whole numbers. In this case they should all have been multiplied by 2.

Anhydrous and hydrated salts

REVISED ☐

Most salts exist in the solid state either as a pure substance or as crystals that have water molecules embedded into their structure. The pure substance is described as **anhydrous,** which means that it contains no water. The crystals containing water are said to possess water of crystallisation and are described as **hydrated**.

The water present in hydrated salts is indicated by writing the formula of the substance followed by a full stop and the number of molecules of water — for example, iron(II) sulfate crystals have the formula $FeSO_4.7H_2O$.

You should be able to calculate the water of crystallisation.

Example 1

A sample of copper sulfate crystals has a mass of 6.80 g. When heated, until all the water of crystallisation has been driven off, the mass is reduced to 4.35 g. Calculate the formula of the copper sulfate crystals.

Answer

mass of water driven off by heating = 6.80 − 4.35 = 2.45 g

moles of water of crystallisation = $\frac{2.45}{18.0}$ = 0.136 mol

molar mass of anhydrous copper sulfate = 63.5 + 32.1 + (4 × 16.0)

= 159.6 g mol⁻¹

moles of copper sulfate = $\frac{4.35}{159.6}$ = 0.02726 mol

mole ratio of copper sulfate to water = 0.02726 : 0.136 or 1:5

Therefore, the formula of hydrated copper sulfate is $CuSO_4.5H_2O$.

Exam tip

The methodology for calculating water of crystallisation is exactly the same as for calculating empirical formula.

Step 1 — find the ratio of the moles of anhydrous salt to moles of water.

Step 2 — divide each by the smallest, and find the simplest whole number ratio.

The ratio should always be 1 mol anhydrous salt:whole number of moles of water.

Example 2

A sample of magnesium sulfate crystals contains 9.78% Mg, 38.69% SO_4^{2-} and 51.53% H_2O by mass. Determine the formula of the crystals.

Answer

$$\text{amount (in moles) of magnesium} = \frac{9.78}{24.3} = 0.4025 \, mol$$

$$\text{amount (in moles) of sulfate} = \frac{38.69}{96.1} = 0.4026 \, mol$$

$$\text{amount (in moles) of water} = \frac{51.53}{18.0} = 2.863 \, mol$$

ratio = 0.4025:0.4026:2.863 or 1:1:7

Therefore, the formula of magnesium sulfate crystals is $MgSO_4.7H_2O$.

Calculating moles from mass

REVISED

$$\text{amount of substance in moles, } n = \frac{\text{mass of substance in grams } (m)}{\text{molar mass of substance } (M)}$$

Therefore,

$$n = \frac{m}{M}$$

This applies to atoms and to compounds. The examples below illustrate the sort of questions that you may be asked. The triangle can be used to help you rearrange the equation.

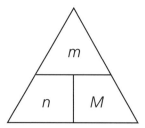

Example 1

Calculate the amount in moles present in 1.49 g of Li_2O.

Answer

mass, m (in grams) = 1.49 g

Using the formula Li_2O, it is possible to deduce the molar mass of Li_2O. Molar mass (M) is calculated by adding together the relative atomic mass of each individual element in the compound.

A_r data: Li = 6.9; O = 16.0

molar mass, M = 6.9 + 6.9 + 16.0 = 29.8

$$n = \frac{m}{M}$$

$$n = \frac{1.49}{29.8} = 0.050 \, mol$$

Example 2

Calculate the mass of 0.25 moles of $NiSO_4$.

Answer

number of moles, n = 0.25 mol

Using the formula $NiSO_4$, it is possible to deduce the molar mass of $NiSO_4$. Molar mass is calculated by adding together the relative atomic mass of each individual element in the compound.

A_r data: Ni = 58.7; S =32.1; O = 16.0

Note that there are four oxygen atoms (4 × 16.0 = 64.0).

molar mass, M, of $NiSO_4$ = 58.7 + 32.1 + 64.0 = 154.8

$$n = \frac{m}{M}$$

Rearranging gives:

$m = nM$

$m = 0.25 × 154.8 = 38.7\,g$

Now test yourself

TESTED

11 Determine the amount, in moles, present in each of the following:
   (a) 8.0 g of sulfur
   (b) 1.68 g of calcium oxide
12 Calculate the mass in grams of each of the following:
   (a) 0.04 mol of aluminium chloride
   (b) 0.45 mol of aluminium hydroxide
13 Calculate the molar mass of an element when 2.60 g is equivalent to 0.05 mol of that element. Identify the element.
14 A compound contains a group 2 metal, X. 0.02 mol of the hydroxide of X has a mass of 2.432 g. Identify the metal X. Show all your working.

Answers on p. 114

Calculating moles from gases

REVISED

It is difficult to measure the mass of a gas but easy to measure the volume. Avogadro deduced that all gases, under the same conditions of temperature and pressure, occupy the same volume and that at room temperature and pressure the volume of 1 mol of a gas is equal to 24 dm³. It follows that the amount in moles of a gas can be calculated using:

$$n = \frac{V\,(\text{in dm}^3)}{24} \qquad \text{or} \qquad n = \frac{V\,(\text{in cm}^3)}{24000}$$

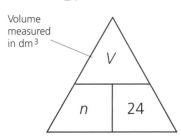
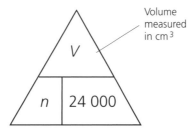

Example

Calculate the amount in moles present in 120 cm³ of hydrogen at room temperature and pressure.

Answer

The volume of gas is given in cm³, so use $n = \dfrac{V}{24000}$.

$$n = \frac{120}{24000} = 0.0050 = 5.0 × 10^{-3}$$

Now test yourself

TESTED

15 When heated, dinitrogen monoxide decomposes to nitrogen and oxygen. The equation for the reaction is:

$$2N_2O(g) \rightarrow 2N_2(g) + O_2(g)$$

What volumes of oxygen and nitrogen are obtained when 50 cm³ of dinitrogen monoxide decomposes?

16 Calculate the mass of each of the following volumes (1 mol of gas has a volume of 24 dm³).
   (a) 4 dm³ of carbon dioxide
   (b) 500 cm³ of ethane

Answers on p. 114

Revision activity

Calculate the approximate volume of you bedroom in cm³. Assume that your room contains 80% nitrogen and 20% oxygen. Calculate the mass of gas in your bedroom.

Calculating moles from solutions

REVISED

Many reactions are carried out in solution. The amount of chemical present is best described by using concentration, c, of the solution in mol dm⁻³ and the volume, V, of the solution.

The units of concentration are nearly always measured in mol dm⁻³; it follows that V is the volume of the solution measured in dm³.

Amount (in moles) in a solution can be calculated using:

$$n = c \times V$$

If V is given in cm³ then use:

$$n = c \times \frac{V(cm^3)}{1000}$$

The examples below illustrate the sort of questions that you may be asked.

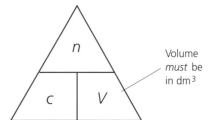

Volume *must* be in dm³

Example 1

Calculate the amount in moles present in 250 cm³ of a 0.50 mol dm⁻³ solution.

Answer

concentration, $c = 0.50$ mol dm⁻³

volume, $V = 250$ cm³

The volume must be converted into dm³:

$$V = \frac{250}{1000} = 0.250 \text{ dm}^3$$

$$n = cV$$

So,

$$n = 0.50 \times 0.250 = 0.125 \text{ mol}$$

Exam practice answers and quick quizzes at **www.hoddereducation.co.uk/myrevisionnotes**

Example 2

Calculate the concentration of a NaOH solution when 4.0 g NaOH is dissolved in 250 cm³.

Answer

This is slightly more complicated. You first have to calculate the amount in moles of NaOH used by using the mass (4.0 g) and the formula, NaOH to deduce the molar mass (23.0 + 16.0 + 1.0 = 40.0):

$$n = \frac{m}{M} = \frac{4.0}{40.0} = 0.10 \, \text{mol}$$

Concentration is calculated thus:

$$c = \frac{n}{V} = \frac{0.10}{250\big/1000} = \frac{0.10}{0.250} = 0.40 \, \text{mol dm}^{-3}$$

Mole calculations appear on almost every exam paper. It is important that you are able to:

- write and balance full equations
- calculate reacting masses
- calculate reacting gas volumes
- calculate reacting volumes of solutions
- calculate concentrations from titrations

The ideal gas equation

REVISED

The particles that make up a gas move in all directions at great speeds. They are so widely spaced that any forces of attraction between the gas particles are insignificant. Table 2.5 compares gases with other states of matter.

Table 2.5

States of matter	Solid	Liquid	Gas
Movement of particles	Vibrate about a fixed position	Slow random movement	Fast random movement
Packing of particles	Close packed	Close packed	Widely spaced
Volume	Fixed	Fixed	Not fixed — the volume changes if temperature or pressure changes
Shape	Fixed	Not fixed — takes on shape of container	Not fixed — takes on shape of container

If the temperature of a gas increases so does the volume, but if the pressure on the gas increases the volume will decrease. The relationship between the volume of a gas and the temperature and the pressure can be expressed in the form of an equation known as the ideal gas equation:

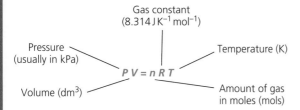

For a given amount of gas:

$$\frac{PV}{T} = \text{constant}$$

The value of the constant is nR, where n is the amount of the gas in moles and R is the gas constant.

It follows that:

$$\frac{P_1 V_1}{T_1} = \frac{P_2 V_2}{T_2}$$

P_1, V_1, and T_1 are the initial pressure, volume and temperature, and P_2, V_2 and T_2 are the final pressure, volume and temperature.

Real gases deviate from ideal gases, but real gases approach ideal behaviour at very high temperatures and at very low pressures.

Exam tip

The value and units of the gas constant will be on the data sheet, which will be supplied in all examinations.

Example 1

If the volume of a gas collected at 50°C and 110 kPa is 75 cm^3, what would the volume be at STP?

For calculations of this type it is best to use the equation:

$$\frac{P_1 V_1}{T_1} = \frac{P_2 V_2}{T_2}$$

The only unknown is V_2 so the equation can be rearranged to give:

$$\frac{P_1 V_1 T_2}{P_2 T_1} = V_2$$

Pressure is in kPa. Standard pressure is 101 kPa.

Temperatures must be in K:

$T_1 = 50°C + 273 = 323\,K$

standard temperature = 0°C = 273 K

The units of volume used for V_1 will also be the units of the calculated volume V_2.

$$\frac{110 \times 75 \times 273}{101 \times 323} = \frac{2\,252\,250}{32\,623} = 69\,cm^3$$

Example 2

2.63 g of a noble gas occupies a volume of 500 cm^3 at RTP. Identify the noble gas.

Answer

Use the ideal gas equation, $PV = nRT$. Because we are using RTP, the only unknown is the amount in moles, n.

The ideal gas equation can be rearranged to give:

$$n = \frac{PV}{RT}$$

$$n = \frac{101 \times 0.5}{8.314 \times 298} = 0.02 \, \text{mol}$$

We now know that 0.02 mols of the noble gas has a mass = 2.63 g, so the atomic mass of the noble gas is:

$$A_r = \frac{m}{n} = \frac{2.63}{0.02} = 131.5$$

Therefore the noble gas is xenon (atomic mass of xenon = 131.3 g mol^{-1}).

Now test yourself

TESTED

17 At standard pressure and a temperature of 37°C, 3.81 g of a halogen gas occupies a volume of 610 cm^3. Identify the halogen gas.

Answer on p. 114

Stoichiometric relationships in calculations

REVISED

The example below shows how reacting masses, reacting gas volumes and reacting volumes of solutions could all be tested in a single equation.

Example

Zinc reacts with dilute hydrochloric acid to produce zinc chloride and hydrogen gas. Write a balanced equation for this reaction and calculate the mass of zinc required to react with 50 cm^3 of a 0.10 mol dm^{-3} solution of dilute hydrochloric acid. Deduce the volume (cm^3) of hydrogen produced in this reaction and calculate the concentration of the zinc chloride solution that would be formed.

Answer

Equation $Zn + 2HCl \rightarrow ZnCl_2 + H_2$

The numbers in front of each formula in the balanced equation tells us the number of moles used and gives the ratio of the reacting moles, i.e. the mole ratio.

Equation	Zn		2HCl		ZnCl$_2$		H$_2$
Mole ratio	1 mol	:	2 mol	:	1 mol	:	1 mol

Exam tip

It is essential that you learn to read like a chemist. Units are important and give clues as to which equation to use:
- mol dm^{-3} indicates concentration and therefore you are likely to use $n = cV$.
- g mol^{-1} indicates molar mass and therefore you are likely to use $n = m/M$.

You can calculate n for HCl since we are given both c and V:

$$n = 0.10 \times \frac{50}{1000} = 0.0050 = 5.0 \times 10^{-3}$$

You can now work out the number of moles of all the other chemicals in the equation by using the mole ratios in the balanced equation:

Equation	Zn	+	2HCl	→	ZnCl$_2$	+	H$_2$
Mole ratio	1 mol	:	2 mol	:	1 mol	:	1 mol
Actual moles	$\frac{5 \times 10^{-3}}{2}$		5.0×10^{-3}		$\frac{5 \times 10^{-3}}{2}$		$\frac{5 \times 10^{-3}}{2}$
	2.5×10^{-3}				2.5×10^{-3}		2.5×10^{-3}

We can use $n = \frac{m}{M}$ to find the mass of Zn required:

$m = nM$

$m = 2.5 \times 10^{-3} \times 65.4$

$m = 0.16\,g$

We can use $n = cV$ to find the concentration of ZnCl$_2$(aq):

$c = \frac{n}{V}$

$c = \frac{2.5 \times 10^{-3}}{50/1000}$

$c = 0.050\,mol\,dm^{-3}$

We can use $n = \frac{V}{24\,000}$ to find the volume of H$_2$ produced:

$V = n \times 24\,000$

$V = 2.5 \times 10^{-3} \times 24\,000$

$V = 60\,cm^3$

Now test yourself

TESTED

18 How many moles are there in each of the following?
   (a) 200 cm³ of 0.5 mol dm⁻³ sulfuric acid
   (b) 25 cm³ of 0.1 mol dm⁻³ hydrochloric acid
19 Calculate the mass of solid sodium hydroxide which, when dissolved, would make 250 cm³ of 0.2 mol dm⁻³ solution of sodium hydroxide.
20 What is the concentration obtained by diluting 50 cm³ of 2 mol dm⁻³ sodium carbonate with water to make:
   (a) 500 cm³ of solution
   (b) 250 cm³ of solution
21 A solution of potassium carbonate has a concentration of 0.04 mol dm⁻³. 25.0 cm³ of this solution is neutralised by 28.10 cm³ of hydrochloric acid.
   (a) Write a balanced equation for the reaction between potassium carbonate and hydrochloric acid.
   (b) How many moles of potassium carbonate are contained in 25.0 cm³ of solution?
   (c) How many moles of hydrochloric acid are needed to neutralise the number of moles of potassium carbonate in 25.0 cm³?
   (d) What is the concentration of the hydrochloric acid in mol dm⁻³?

Answers on pp. 114–115

Percentage yield calculations

Percentage yield calculations involve mole calculations and are often used when preparing organic compounds. Reactions of organic molecules will be covered fully in Module 4 (AS) and Module 6 (A level). A typical calculation is shown below.

Example

Ethanol, C_2H_5OH, can be oxidised to form ethanal, CH_3CHO. If 2.30 g of ethanol are oxidised to produce 1.32 g of ethanal, calculate the percentage yield.

Answer

Any mole calculations require a balanced equation, so it is essential that you are able to write suitable balanced equations and to use the mole ratios from the equation. [O] can be used to represent the oxidising agent.

Equation	C_2H_5OH	+	[O]	$\rightarrow$	CH_3CHO	+	H_2O
Mole ratio	1 mol		1 mol		1 mol		1 mol

The equation shows that 1 mole of ethanol produces 1 mole of ethanal.

Step 1: calculate the number of moles of ethanol used:

$$\text{amount of ethanol used} = n = \frac{\text{mass (in g)}}{\text{molar mass}}$$

$$n = \frac{m}{M}$$

$$\text{amount of ethanol used} = n = \frac{2.3}{46} = 0.05 \text{ mol}$$

Since the mole ratio is 1:1, the amount of ethanal that could be made is also 0.05 mol.

Step 2: Calculate the number of moles of ethanal actually produced:

$$\text{amount of ethanal produced} = n = \frac{m}{M}$$

$$= \frac{1.32}{44} = 0.03 \text{ mol}$$

Step 3: Calculate the percentage yield by using:

$$\text{percentage yield} = \frac{\text{actual yield} \times 100}{\text{maximum yield}}$$

$$= 0.03 \times \frac{100}{0.05}$$

$$= 60\%$$

Now test yourself

22 A student reacted 4.60 g of ethanol (C_2H_5OH) with an excess of methanoic acid (HCOOH) and made 5.92 g of ethyl methanoate ($HCOOC_2H_5$). Calculate the student's percentage yield. The equation for the reaction is:

$$C_2H_5OH + HCOOH \rightleftharpoons HCOOC_2H_5 + H_2O$$

Answer on p. 115

Atom economy calculations

$$\text{atom economy} = \frac{\text{molar mass of desired product}}{\text{molar mass of all products}} \times 100$$

The reaction between ethanol, C_2H_5OH, and ethanoic acid, CH_3COOH, is used to make the ester ethyl ethanoate, $CH_3COOC_2H_5$.

$$CH_3COOH + C_2H_5OH \rightarrow CH_3COOC_2H_5 + H_2O$$

The desired product is the ester, but water is also produced.

molar mass of $CH_3COOC_2H_5$ = 88 g

molar mass of H_2O = 18 g

$$\text{atom economy} = \frac{88}{88 + 18} \times 100$$

$$= \frac{88}{106} \times 100 = 83\%$$

A high atom economy is good, since that would indicate a low level of waste. Atom economy can be greatly enhanced if a use can be found for the by-product.

> **Exam tip**
>
> If asked to define atom economy, simply use this equation.

> **Exam tip**
>
> When carrying out calculations do not round numbers during the calculation. Keep the number in your calculator and only round when you have finished the entire calculation.

Now test yourself

23 A student prepared a sample of methanol, CH_3OH, by reacting bromomethane with potassium hydroxide.

$$CH_3Br + KOH \rightarrow CH_3OH + KBr$$

Calculate the atom economy for the preparation of methanol.

Answer on p. 115

Acids

Acids and bases

The release of a proton, H^+, by an **acid** can only occur in aqueous solution. Pure HCl is a covalent gas and it is only when it comes into contact with water that it can release a H^+ (proton):

$$HCl(g) + H_2O(l) \rightarrow H_3O^+(aq) + Cl^-(aq)$$

$H_3O^+(aq)$ is usually written as $H^+(aq)$ and the above equation shown as:

$$HCl(aq) \rightarrow H^+(aq) + Cl^-(aq)$$

> An **acid** is a proton donor.

You are expected to know the formulae of some common acids: hydrochloric acid, HCl; sulfuric acid, H_2SO_4; nitric acid, HNO_3; and ethanoic acid, CH_3COOH. The acidic proton in ethanoic acid is $CH_3COO\mathbf{H}$.

Acids have a pH below 7, and the stronger acids have lower pHs. Acids can be sub-divided into strong acids and weak acids.

● Strong acids include HCl, H_2SO_4 and HNO_3. They are regarded as strong acids because in solution they totally dissociate into their ions. The equation always includes the '$\rightarrow$' symbol.

$$HCl(aq) \rightarrow H^+(aq) + Cl^-(aq)$$

- Weak acids include organic acids such as ethanoic acid, CH_3COOH. They are weak acids because in solution they only partially dissociate into their ions. The equation always includes the '$\rightleftharpoons$' symbol.

$$CH_3COOH(aq) \rightleftharpoons CH_3COO^-(aq) + H^+(aq)$$

Common **bases** include metal oxides (e.g. Na_2O, MgO, CuO), metal hydroxides (e.g. $NaOH$, KOH, $Cu(OH)_2$), ammonium hydroxide (NH_4OH) and metal carbonates (e.g. K_2CO_3, $ZnCO_3$). Hydroxides, such as $NaOH(aq)$, are strong bases and completely dissociate, whereas $NH_4OH(aq)$ is a weak base and only partially dissociates:

$$NaOH(aq) \rightarrow Na^+(aq) + OH^-(aq)$$

$$NH_4OH(aq) \rightleftharpoons NH_4^+(aq) + OH^-(aq)$$

Common **alkalis** include group 1 hydroxides, such as $LiOH$ and $NaOH$, and some group 2 hydroxides, for example $Ca(OH)_2$.

Salts

REVISED

A **salt** is formed when an acid reacts with a metal, a carbonate, a base or an alkali, as in the following examples. Acids react in aqueous solution and all reactions can be represented with either a balanced equation or an ionic equation:

Acid and metal:

Balanced equation	$2HCl(aq) + Mg(s) \rightarrow MgCl_2(aq) + H_2(g)$
Ionic equation	$2H^+(aq) + Mg(s) \rightarrow Mg^{2+}(aq) + H_2(g)$

Acid and carbonate:

Balanced equation	$2HCl(aq) + K_2CO_3(aq) \rightarrow 2KCl(aq) + H_2O(l) + CO_2(g)$
Ionic equation	$2H^+(aq) + CO_3^{2-}(aq) \rightarrow H_2O(l) + CO_2(g)$

Acid and base:

Balanced equation	$2HCl(aq) + CuO(s) \rightarrow CuCl_2(aq) + H_2O(l)$
Ionic equation	$2H^+(aq) + CuO(s) \rightarrow Cu^{2+}(aq) + H_2O(l)$

Acid and an alkali:

Balanced equation	$HCl(aq) + NaOH(aq) \rightarrow NaCl(aq) + H_2O(l)$
Ionic equation	$H^+(aq) + OH^-(aq) \rightarrow H_2O(l)$

Bases such as ammonia react with acids to produce a salt. Ammonium sulfate, $(NH_4)_2SO_4$, is used as a fertiliser and is manufactured by reacting ammonia with sulfuric acid:

Balanced equation	$2NH_3(aq) + H_2SO_4(aq) \rightarrow (NH_4)_2SO_4(aq)$

You are expected to do acid–base titrations in the laboratory and then carry out calculations. A titration is a process whereby a precise volume of one solution is added to another solution until the exact volume required to complete the reaction has been found. You are expected to obtain results accurate to within $0.10\,cm^3$. Practical skills will be examined in all exams.

Revision activity

There are very many naturally occurring organic acids. Find the formulae of: lactic acid, malic acid, tartaric acid, ascorbic acid and citric acid. What do they all have in common?

A **base** is a proton acceptor.

An **alkali** is a soluble base that releases hydroxide ions, OH^- when in aqueous solution.

A **salt** is formed when an acid has one or more of its hydrogen ions replaced by either a metal ion or an ammonium ion, NH_4^+.

Typical mistakes

Examiners often use unfamiliar acids such as phosphoric acid, H_3PO_4, and expect students to be able to work out the formulae of salts formed from the acid. Statistics show that the majority of students find this difficult. The way to work it out is to use the clue in H_3PO_4. The phosphate ion combines with three hydrogens, so it must be valency 3. Therefore, if it forms a salt with a group 1 metal, for example sodium, the formula is Na_3PO_4; with a group 2 metal, for example magnesium, the formula is $Mg_3(PO_4)_2$.

Exam tip

Organic acids such as ethanoic acid contain –COOH, and when they react the H in the –COOH is replaced by a metal ion or an ammonium ion.

Example

It was found that $18.60\,cm^3$ HCl(aq) neutralised exactly $25.0\,cm^3$ $0.100\,mol\,dm^{-3}$ NaOH(aq). Calculate the concentration of the HCl(aq) solution.

Answer

Step 1 Work out how many moles, n, of sodium hydroxide were used. This is possible because the concentration, c, and the volume, V, are known.

The concentration of the sodium hydroxide is $0.100\,mol\,dm^{-3}$ of solution and $25.0\,cm^3$ was used.

$$n = cV = 0.100 \times \frac{25.0}{1000} = 0.00250\,mol$$

Step 2 Refer to the balanced equation to see how many moles of hydrochloric acid are needed to react with this number of moles of sodium hydroxide.

The equation for the reaction is: $\quad$ NaOH $\quad + \quad$ HCl $\quad \rightarrow \quad$ NaCl $\quad + \quad H_2O$

The mole ratio is: $\qquad\qquad\qquad\qquad$ 1 $\quad : \quad$ 1 $\quad : \quad$ 1 $\quad : \quad$ 1

Therefore, 1 mol of NaOH reacts with 1 mol of HCl.

As $0.00250\,mol$ of NaOH were used, $0.00250\,mol$ of HCl must be required to react completely.

Since $18.60\,cm^3$ of HCl were added from the burette, $18.60\,cm^3$ of HCl must contain $0.00250\,mol$.

Step 3 Convert the information obtained about the hydrochloric acid into its concentration in $mol\,dm^{-3}$.

$$c = \frac{n}{v} = \frac{0.00250}{18.6/1000} = \frac{0.00250}{0.0186} = 0.134\,mol\,dm^{-3}$$

Exam tip

All ions have the state symbol (aq). When balancing an ionic equation, you must balance charge as well as symbols.

Now test yourself

TESTED ☐

24 Write the formulae of the following salts.
  (a) calcium nitrate
  (b) aluminium sulfate
  (c) magnesium ethanoate
25 Write full and ionic equations for each of the following reactions:
  (a) aqueous solutions of ethanoic acid and sodium hydroxide
  (b) solid $CaCO_3$ and aqueous nitric acid

Answers on p. 115

Redox

A species is said to be **oxidised** if it loses electrons. The converse is true for **reduction**. One way of remembering this is the acronym OILRIG:

Oxidation **I**s **L**oss **R**eduction **I**s **G**ain

Oxidation involves the loss of electrons or an increase in the oxidation number.

Reduction involves the gain of electrons or a decrease in the oxidation number.

Redox reactions are reactions in which electrons are transferred from one substance to another.

Oxidation number

Oxidation number is a convenient way of identifying whether or not a substance has undergone either **oxidation** or **reduction**. In order to work out the oxidation number you must learn a few simple rules (Table 2.6).

Table 2.6 Rules for determining oxidation numbers

Rule	Example
All elements in their natural state have the oxidation number zero	Hydrogen, H_2; oxidation number = 0
Oxidation numbers of the atoms of any compound always add up to zero	Carbon dioxide, CO_2; sum of oxidation numbers = 0
Oxidation numbers of the components of any ion always add up to the charge of the ion	Nitrate, NO_3^-; sum of oxidation numbers = –1

There are certain elements whose oxidation numbers never change but some other elements have variable oxidation numbers and these have to be deduced.

When calculating the oxidation numbers of elements in either a compound or an ion you should apply the following order of priority:
1 The oxidation numbers of elements in groups 1, 2 and 3 are always +1, +2 and +3 respectively.
2 The oxidation number of fluorine is always −1.
3 The oxidation number of hydrogen is usually +1.
4 The oxidation number of oxygen is usually −2.
5 The oxidation number of chlorine is usually −1.

By applying these rules *in sequence*, it is possible to deduce any oxidation number.

Example 1

Deduce the oxidation number of Mn in $KMnO_4$.

Answer

$KMnO_4$ is a compound and therefore the oxidation numbers must add up to zero.

In order of priority K comes first and its oxidation number is +1; O is second and its oxidation number is –2 *but* there are 4 oxygens hence a total of –8.

In order for the oxidation numbers to add up to zero, the oxidation number of Mn in $KMnO_4$ must be +7.

Example 2

Deduce the oxidation number of I in IO_3^-.

Answer

IO_3^- is an ion and therefore the oxidation numbers must add up to the charge on the ion, i.e. they must add up to –1.

In order of priority, O comes first and its oxidation number is –2. However, there are three oxygens in IO_3^-, so the total for oxygen is –6.

In order for the oxidation numbers to add up to the charge of the ion (–1) the oxidation number of I in IO_3^- must be +5.

When magnesium reacts with steam, magnesium oxide and hydrogen are formed:

$$Mg(s) + H_2O(g) \rightarrow MgO(s) + H_2(g)$$

It is easy to see that magnesium has been oxidised (it has gained oxygen) and that water has been reduced (it has lost oxygen). The oxidation numbers for this reaction are shown in Figure 2.3.

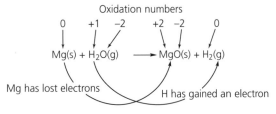

Figure 2.3

An increase in oxidation number is due to oxidation.

A decrease in oxidation number is due to reduction.

Electron transfer can be shown by using ionic half-equations, for example:

$Mg \rightarrow Mg^{2+} + 2e^-$ (*loss* of electrons = oxidation)

$2H^+ + 2e^- \rightarrow H_2$ (*gain* of electrons = reduction)

Now test yourself

26 Deduce the oxidation numbers in each of the following:
   (a) H_2O, $NaOH$, KNO_3, NH_3, N_2O
   (b) SO_4^{2-}, CO_3^{2-}, NH_4^+, MnO_4^-, $Cr_2O_7^{2-}$

Answer on p. 115

Redox reactions

Metals generally react by losing electrons (oxidation is loss), so the oxidation number increases, for example:

	Mg	$\rightarrow$	Mg^{2+}	+	$2e^-$
Oxidation numbers:	0		+2		

Non-metals generally react by gaining electrons (reduction is gain), so the oxidation number decreases, for example:

	F_2	+	$2e^-$	$\rightarrow$	$2F^-$
Oxidation numbers:	0				−1

Exam tip

Redox reactions occur throughout the A-level course, so it is essential that you can work out oxidation numbers. This becomes more important in the second year of the course when studying redox titrations and electrode potentials. Make sure that you know the rules and can apply the order of priority for working out oxidation numbers: groups 1, 2 and 3, followed by F, H, O and Cl.

Now test yourself

27 Use oxidation numbers to identify what has been oxidised in the following reaction:

$$Zn + CuSO_4 \rightarrow Cu + ZnSO_4$$

Answer on p. 115

Exam practice

1 (a) A chemist reacted oxygen separately with magnesium and with sulfur to form magnesium oxide and sulfur dioxide respectively. Write an equation for each reaction. [1]
  (b) The reactions in (a) are both redox reactions in which reduction and oxidation take place. Explain, using the changes in oxidation number for sulfur, whether sulfur has undergone oxidation or reduction. [2]
  (c) The chemist added water to magnesium oxide and to sulfur dioxide, forming two aqueous solutions. Write equations for the reactions that took place. [2]

2 10.00 g of sodium carbonate crystals $Na_2CO_3.xH_2O$ are dissolved to make 1.00 dm³ of solution. 25.0 cm³ of this solution are neutralised by 17.50 cm³ of 0.100 mol dm⁻³ hydrochloric acid.
  (a) Calculate the amount, in moles, of hydrochloric acid present in 17.50 cm³. [1]
  (b) Deduce the number of moles of sodium carbonate present in 25.0 cm³. [2]
  (c) Calculate the number of moles and mass of sodium carbonate in 1.0 dm⁻³ of the solution. [2]
  (d) Use your answer to (c) to deduce the mass of water in the crystals. [2]
  (e) Deduce the value of x in $Na_2CO_3.xH_2O$. [2]

Answers and quick quiz 2 online

ONLINE ☐

Summary

You should now have an understanding of:
- atomic structure, isotopes, relative masses, mass spectra
- how to deduce formulae and write equations
- mole calculations using the equations: $n = m/M$, $n = V/24$ and $n = cV$
- the ideal gas equation

- acids, bases and the formation of salts and be able to calculate the formulae of hydrated salts
- titrations
- oxidation and reduction in terms of oxidation numbers

3 Electrons, bonding and structure

Electron structure

Ionisation energy

REVISED

Ionisation energy gives evidence for the existence of shells and subshells of electrons.

The **first ionisation energy** can be represented by the equation:

$$X(g) \rightarrow X^+(g) + e^-$$

It is important to include the state symbols (g).

For elements that have more than one electron, it is possible to remove each electron stepwise.

The **second ionisation energy** is represented by the equation:

$$X^+(g) \rightarrow X^{2+}(g) + e^-$$

The second ionisation energy results in the formation of a 2+ ion and starts with a 1+ ion. It follows that the **nth ionisation energy** is:

$$X^{(n-1)+}(g) \rightarrow X^{n+}(g) + e^-$$

> **Typical mistakes**
>
> When asked to write an equation to illustrate the third ionisation energy many students write: $X(g) \rightarrow X^{3+}(g) + 3e^-$. This is incorrect as it represents the first, second and third ionisation energies combined.
>
> The correct response is $X^{2+}(g) \rightarrow X^{3+}(g) + e^-$. It is worth remembering that there is always just one e^- on the right-hand side of the equation and that the charges on both sides have to balance.

> The **first ionisation energy** of an element is the energy required to remove one electron from each atom in one mole of gaseous atoms to form one mole of gaseous ions of charge 1+.

> The **second ionisation energy** is the energy required to remove one electron from each ion in one mole of gaseous ions of charge 1+ to form one mole of gaseous ions of charge 2+.

> The **nth ionisation energy** is the energy required to remove one electron from each ion in one mole of gaseous ions of charge $(n-1)^+$ ions to form one mole of gaseous n^+ ions.

Trends in ionisation energies

REVISED

There are *three* factors that influence ionisation energy.

Factor 1 — distance of the outermost electron from the nucleus (atomic radius)

Factor 2 — electron shielding (the number of inner shells)

Factor 3 — nuclear charge (the number of protons in the nucleus)

Down a group:

Factor 1 — atomic radii increases, which should make it easier to remove an electron

Factor 2 — shielding increases, which should make it easier to remove an electron

Factor 3 — nuclear charge increases, which should make it more difficult to remove an electron

Factors 1 and 2 outweigh factor 3, so ionisation energy *decreases* down a group.

Across a period:

Factor 1 — atomic radii decreases, which should make it more difficult to remove an electron

Factor 2 — shielding remains the same, which has no effect on the ease of removing an electron

Factor 3 — nuclear charge increases, which should make it more difficult to remove an electron

Factors 1 and 3 mean that ionisation energy *increases* across a period.

It is possible to remove electrons stepwise from an atom and to measure the size of successive ionisation energies. When the successive energies are plotted, the graph provides evidence for the existence of shells. Phosphorus has 15 electrons, which are arranged 2,8,5. The evidence for this is the plot of successive ionisation energies shown (Figure 3.1).

Exam tip

This is how to sketch a plot of successive ionisation energies of an element — for example $_{13}$Al. The atomic number 13 tells you that aluminium has 13 protons and 13 electrons. Use your GCSE knowledge to work out that the electrons are arranged 2, 8, 3. When sketching the successive ionisation energies, remember that the electrons in the outer shell are removed first. In this case, therefore, you draw the ionisation energies of the three outer shell electrons followed by a big increase, then the ionisation energies for the next shell, which contains eight electrons, followed by another big jump, and finally the ionisation energies for the inner two electrons.

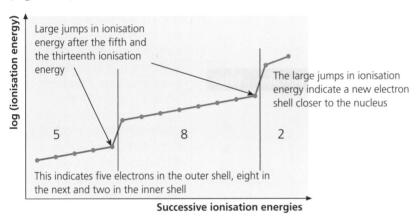

Figure 3.1 Successive ionisation energies of $_{15}$P

Plotting successive ionisation energies confirms what you learnt at GCSE. It shows that the first shell contains a maximum of two electrons and the second shell contains a maximum of eight electrons.

There is further experimental evidence to suggest that each shell is made of smaller subshells.

Now test yourself

TESTED

1 Element X is in period 3. The first seven successive ionisation energies are shown below:
1012, 1903, 2912, 4957, 6274, 21 269, 25 398
   (a) Use these ionisation energies to identify element X. Explain your reasoning.
   (b) Write an equation to represent the third ionisation energy of element X.

Answer on p. 115

By studying the first ionisation energies of the first 20 elements we obtain evidence for the existence of subshells (Figure 3.2).

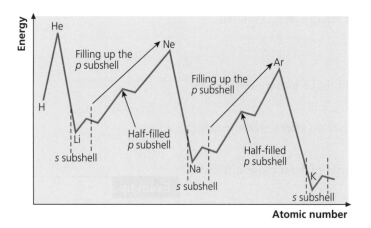

Figure 3.2 Evidence for the existence of subshells

The graph shows that there is a gradual increase in ionisation energies across a period, and it also reveals several small peaks and troughs. These peaks and troughs are repeated in each period. They provide evidence for the existence of subshells (*s*, *p* and *d*). There is a periodic variation throughout all of the elements and there is evidence for several shells and subshells.

Table 3.1 Shells and subshells

Shell	Subshells				Total number of electrons				Total
1st	1s				2				2
2nd	2s	2p			2	6			8
3rd	3s	3p	3d		2	6	10		18
4th	4s	4p	4d	4f	2	6	10	14	32

We now know that the subshells are made up of orbitals.

> **Exam tip**
>
> The specification does not require you to explain why the ionisation energy drops immediately after a half-filled *p* subshell.

> **Typical mistake**
>
> Questions often ask students to complete or to state the number of electrons in a *p*-orbital or a *d* subshell or the third shell and it is essential to differentiate between *orbitals*, *subshells and shells*. All orbitals contain a maximum of two electrons, subshells are made up orbitals and shells consist of subshells. Table 3.1 demonstrates this.

Now test yourself

TESTED

2 Explain why the first ionisation energy of potassium is less than that of sodium.

Answer on p. 115

Electron configuration

REVISED

The concept of an orbital is difficult and if you are asked to define an orbital, the simplest definition is:

An orbital is a region around the nucleus that can hold up to two electrons with opposite spin.

You should be able to describe, with the aid of a diagram, the shape of the *s*- and the *p*-orbitals.

There are three *p*-orbitals, one along each of the *x*, *y* and *z* axes (Figure 3.3).

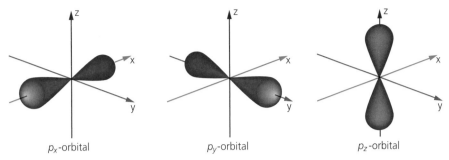

p_x-orbital
p_y-orbital
p_z-orbital

Figure 3.3

The sequence in which electrons fill the orbitals is shown in Figure 3.4.

Remember that, within an orbital, the electrons have opposite spins. The lowest energy level is occupied first; orbitals at the same energy level are occupied singly before pairing of electrons.

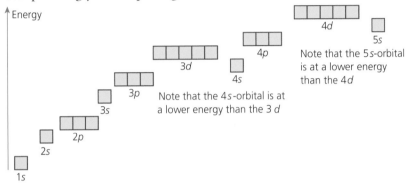

Figure 3.4 Sequence in which the orbitals are filled by electrons

You should be able to write the full electron configuration of the first 36 elements. For example:

$_{15}$P is $1s^2 2s^2 2p^6 3s^2 3p^3$ or $[_{10}Ne] \, 3s^2 3p^3$

$_{25}$Mn is $1s^2 2s^2 2p^6 3s^2 3p^6 3d^5 4s^2$ or $[_{18}Ar] \, 3d^5 4s^2$

Exam tip

Most students remember that the 4s fills before the 3d. However, when metals lose electrons to form ions, the 4s is also lost before the 3d. So, for example, the electron configuration of $_{22}$Ti^{2+} is $1s^2 2s^2 2p^6 3s^2 3p^6 3d^2$ and *not* $1s^2 2s^2 2p^6 3s^2 3p^6 4s^2$ (the Ti atom is $1s^2 2s^2 2p^6 3s^2 3p^6 3d^2 4s^2$).

Now test yourself

TESTED

3 Write the electron configuration of each of: N, Al^{3+}, P^{3-}, Fe^{3+}.

Answer on p. 115

Bonding and structure

Types of bond

REVISED

There are three main types of bond: **ionic**, **covalent** and **metallic**. Table 3.2 summarises each type.

An **ionic bond** is the electrostatic attraction between oppositely charged ions.

A **covalent bond** is the strong electrostatic attraction between a shared pair of electrons and the nuclei of the bonded atoms. Each atom provides one of the shared pair.

A **metallic bond** is the electrostatic attraction between positive metal ions in the lattice and the delocalised electrons.

Table 3.2 Properties of the three types of bond

	Ionic	Covalent	Metallic
Formation	Formed by electron transfer from metal atom (X) to non-metal atom (Y) to produce oppositely charged ions X$^+$ and Y$^-$	Formed when electrons are shared rather than transferred	The positive ions occupy fixed positions in a lattice and the delocalised electrons can move freely throughout the lattice
Direction	An ionic bond is directional acting between adjacent ions	A covalent bond is directional, acting solely between the two atoms involved in the bond	A metallic bond is non-directional because the delocalised electrons can move anywhere in the lattice
Examples	KBr, CuO	O_2, C_2H_6	Ni, Mg
Melting and boiling points	High melting point and boiling point due to strong electrostatic forces between ions throughout the lattice	Low melting point and boiling point due to the simple molecular structure being held together by weak forces between molecules	High melting point and boiling point due to strong metallic bonds between the positive ions and negative electrons throughout the lattice
Conductivity	Non-conductor of electricity in solid state, but conducts when molten or dissolved in water because the ionic lattice breaks down and ions are free to move as mobile charge carriers	Non-conductors of electricity — no free or mobile charged particles	Good thermal and electrical conductors due to mobile, delocalised electrons that conduct heat and electricity, even in the solid state
Solubility	The ionic lattice dissolves in polar solvents (e.g. water) because polar water molecules attract ions in the lattice and surround each ion (hydration)	Simple molecular structures soluble in non-polar solvents (e.g. hexane) but usually insoluble in water	Insoluble in polar and non-polar solvents Some metals react with water

In addition to covalent bonding, **dative covalent** or coordinate bonds exist. These are also the result of two shared electrons but in this case one of the atoms supplies both shared electrons.

Typical mistake

When asked to describe bonding and properties students often lose marks by careless use of technical terms — for example, by describing an ionic bond as the attraction between oppositely charged atoms, rather than between oppositely charged ions.

Revision exercise

Look up the conductivity of aluminium, copper, silver, gold, platinum. Put them in order of conductivity. Which is the best conductor? Why is the best conductor not used for electrical wiring?

Dot-and-cross diagrams

REVISED

Dot-and-cross diagrams are a simple visual way to illustrate both ionic and covalent bonding Figure 3.5.

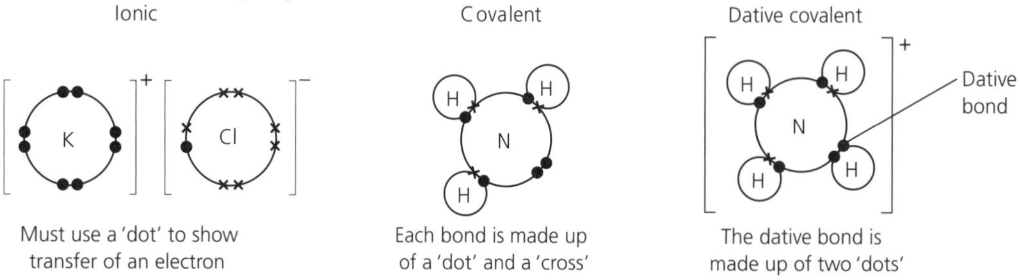

Figure 3.5 Bonding in potassium chloride, ammonia and the ammonium ion

Typical mistake

Exam questions may ask students to draw a dot-and-cross diagram of a compound such as magnesium oxide, MgO. A common incorrect response is shown in Figure 3.6.

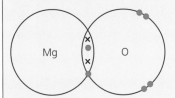

Figure 3.6

Here, magnesium oxide is shown as covalent (sharing electrons), when it is ionic. The two outer electrons of magnesium should be transferred to oxygen, resulting in the formation of Mg^{2+} and O^{2-} ions. A good guide is to look at the mark allocation — 1 mark usually indicates that a covalent structure is required, 2 marks usually indicates ionic.

Shapes of molecules

The main points of the **electron pair repulsion theory** are:
- Electron pairs repel one another as far apart as possible.
- The shape depends upon the number and type of electron pairs surrounding the central atom.
- Lone pairs of electrons are more 'repelling' than bonded pairs of electrons.
- The order of repulsion is: lone pair–lone pair > lone pair–bonded pair > bonded pair–bonded pair.

You should be able to draw a dot-and-cross diagram of a molecule and use it to determine the number and type of electron pairs around the central atom. Then use Table 3.3 to predict the shape, the bond angle and whether or not it is symmetrical.

Table 3.3 Shapes of molecules

Number of bonded pairs of electrons	Number of lone pairs of electrons	Shape	Approximate bond angle	Symmetry
2	0	Linear	180°	Yes
3	0	Trigonal planar	120°	Yes
4	0	Tetrahedral	109.5°	Yes
5	0	Trigonal bipyramidal	90° and 120°	Yes
6	0	Octahedral	90°	Yes
3	1	Pyramidal	107°	No
2	2	Angular	104°	No

The **electron-pair repulsion theory** states that electron pairs repel each other and the shape of a covalent molecule is determined by the number (and type) of electron pairs around the central atom.

Exam tip

In exams, questions asking for an explanation of the electron pair repulsion theory are often answered badly. Learn the key definition above and the bullet points. If you know these, you will score 3 or 4 easy marks.

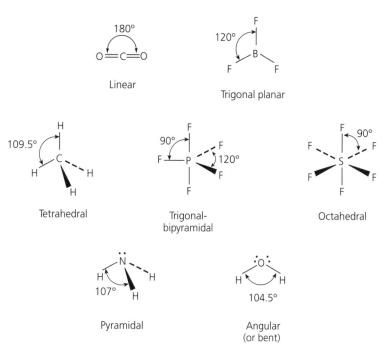

Figure 3.7

Now test yourself

TESTED

4 Determine the shape of each of the following molecules and ions. You may find it helpful to refer to the periodic table to establish the number of outer shell electrons.
   (a) H_2S
   (b) PH_3
   (c) SCl_2
   (d) CH_2Cl_2
   (e) CH_3^+
   (f) H_3O^+

Answers on p. 115

Electronegativity

REVISED

Ionic and covalent bonds are extremes — most bonds are neither 100% ionic nor 100% covalent but somewhere in-between.

When a covalent bond forms between two different elements it is likely that the elements will attract the covalent bonded pair of electrons unequally. The ability of an atom to attract the bonding electrons in a covalent bond is known as the **electronegativity**.

Electronegativity increases across a period but decreases down a group such that fluorine is the most electronegative element.

Electronegativity is the attraction that each bonded atom has for the pair of electrons in the gaseous covalent bond.

Table 3.4 Electronegativity

Li	Be	B	C	N	O	F
1.0	1.5	2.0	2.5	3.0	3.5	4.0
Na						Cl
0.9						3.0
K						Br
0.8						2.8

In molecules like HCl the two electrons in the covalent bond are shared unequally. Chlorine has a higher electronegativity than hydrogen, so the two shared electrons are pulled towards the chlorine atom resulting in the formation of a permanent dipole.

The bond in HCl consists of two shared electrons (essentially covalent) but also contains $\delta+$ and $\delta-$ charges (hence the ionic character).

- The greater the difference between electronegativities, the greater the ionic character of the bond.
- The greater the similarity in electronegativities, the greater the covalent character of the bond.

Bond polarity

In general, a compound made of two or more different non-metals will be a **polar molecule** unless the molecule is symmetrical, in which case any dipoles due to **polar bonds** cancel out.

Common examples include HCl, H_2O and NH_3 (Figure 3.8).

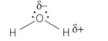

Figure 3.8 Some polar molecules

If a molecule is formed from two different elements the dipoles usually cancel if the molecule is trigonal planar, tetrahedral, trigonal bipyramidal or octahedral. Molecules with these shapes tend to be non-polar.

> A **polar bond** is a covalent bond between two atoms with different electronegativities in which the bonded electron pair is drawn closer to the more electronegative atom.
>
> A **polar molecule** is a molecule in which the electron density is not equally distributed resulting in areas of high electron density ($\delta-$) and areas of low electron density ($\delta+$)

Intermolecular forces

Ionic, covalent and metallic bonds are all strong bonds with bond enthalpies from $200–600\,kJ\,mol^{-1}$. There are three other types of bond that are much weaker, having enthalpies from $2–40\,kJ\,mol^{-1}$. These bonds are formed *between* molecules and are collectively known as **intermolecular** forces.

Permanent dipole–dipole interactions (van der Waals forces) are usually found between polar molecules such as HCl. Dipole–dipole interactions are weak intermolecular forces between the permanent dipoles of different molecules (Figure 3.9).

Induced dipole–dipole interactions (van der Waals forces or London dispersion forces) are the weakest of the forces and act between all molecules, polar or non-polar. They are caused by the movement of electrons. The strength of the induced dipole–dipole interactions depends on the number of electrons in the molecule. The more electrons present in an atom or molecule, the greater are the forces.

If you are asked to explain or define an induced dipole–dipole interaction (van der Waals force) there are three key features that you must include:
- the movement of electrons generates an instantaneous dipole
- this instantaneous dipole induces another dipole in neighbouring atoms or molecules
- the attraction between the temporary induced dipoles results in the dipole–dipole interaction

Weak force of attraction between the $Cl^{\delta-}$ in one HCl and the $H^{\delta+}$ in the next HCl

Figure 3.9

Hydrogen bonds (Figure 3.10) exist between molecules that contain hydrogen atoms bonded to nitrogen, oxygen or fluorine. They are comparatively strong dipole–dipole interactions. Hydrogen bonds exist, for example, between molecules in NH_3, H_2O and HF. The lone pairs of electrons on the nitrogen, oxygen and fluorine play an essential role in the formation of hydrogen bonds. Hydrogen bonds are also found in alcohols.

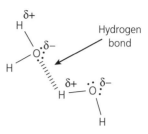

Figure 3.10

Typical mistake

Alcohols, such as methanol (CH_3OH), also form hydrogen bonds. If asked to draw two molecules of methanol and show how hydrogen bonds are formed, students often draw the hydrogen bond in the wrong position, as shown in Figure 3.12a. Figure 3.12b shows the correct response.

Figure 3.12

Exam tip

When asked to show how a hydrogen bond is formed between two water molecules, most students score the marks. However, when asked to show how a hydrogen bond is formed between two ammonia molecules, most students fail to score. You need to practise this. The correct diagram is shown in Figure 3.11.

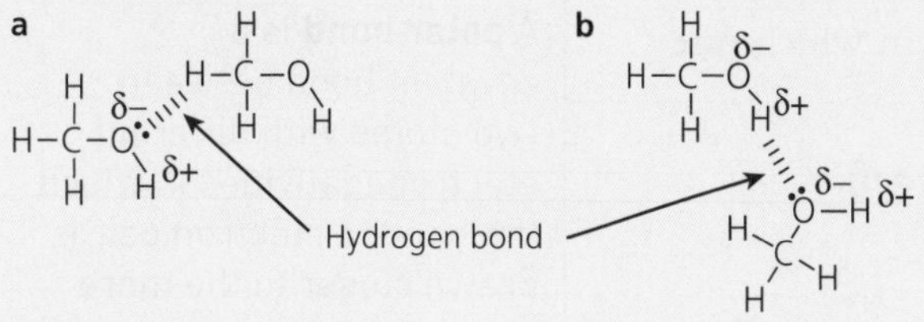

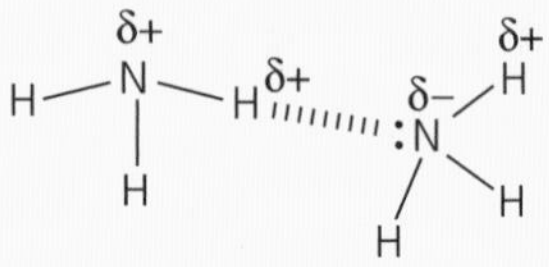

Figure 3.11

Special properties of water arising from hydrogen bonding are:

Solid (ice) is less dense than liquid (water) because the hydrogen bonds in ice hold the H_2O molecules further apart, creating an open lattice structure.

The melting point and boiling point of water are higher than expected owing to the additional energy required to break the hydrogen bonds.

Typical mistake

When asked to explain the special properties of water many students state that water has a high melting point. We all know that water melts at 0°C — which is *not* very high. Water has a higher *than expected* melting point. The evidence for this is found by comparing the melting point of H_2O with those of H_2S, H_2Se and H_2Te, all of which are group 6 hydrides.

Bonding and physical properties

REVISED

Giant ionic structures are held together by strong electrostatic attractions between the ions throughout the lattice. Properties include the following:

- They have high melting points and boiling points.
- They are good conductors when molten or aqueous because they only have mobile charged particles(ions) when molten or when dissolved in water.
- They are soluble in polar solvents such as water.

Simple covalent structures are molecular; the molecules are held together by weak intermolecular forces. However, some covalent molecules, such as water and iodine, also form **simple molecular lattices**. The molecules are held in position in the lattice by intermolecular forces that are comparatively easy to break. Properties include the following:

- They have lower melting points and boiling points.
- They are poor conductors because they do not have any mobile charged particles (electrons or ions).
- They are soluble in non-polar solvents such as water.

Exam practice

1 (a) Explain what is meant by the term *electronegativity*. [2]
   (b) Draw a diagram to show hydrogen bonding between two molecules of water. Your diagram must include the bond angle, the dipoles and relevant lone pairs of electrons. [4]
   (c) State and explain two properties of ice that are a direct result of hydrogen bonding. [4]
2 The electron-pair repulsion theory can be used to predict the shape of covalent molecules. State what is meant by the term *electron pair repulsion theory* and use it to determine the shapes of four molecules of your choice. Choose molecules that illustrate four different shapes. State the bond angle in each shape. [11]
3 Magnesium oxide is a solid with melting point 2852°C; the melting point of sulfur dioxide is −73°C. Explain, in terms of structure and bonding, why there is such a large difference between the melting points of these two oxides. [6]
4 Chlorine reacts with sodium to form sodium chloride.
   (a) Describe the bonding in Cl_2, Na and NaCl. [8]
   (b) Relate the physical properties of Cl_2 and NaCl to their structure and bonding. [8]

Answers and quick quiz 3 online

ONLINE

Summary

You should now have an understanding of:
- ionisation energies
- electron configuration using $1s^22s^2$... notation
- ionic, covalent and metallic bonding
- shapes of covalent molecules and ions
- electronegativity and bond polarity
- intermolecular forces including hydrogen bonding and van der Waals forces

4 The periodic table

Periodicity

Structure of the periodic table

The periodic table is the arrangement of elements by increasing atomic number. Elements with the same outer shell electron configuration are grouped together, so physical and chemical properties are repeated periodically.

The International Union of Pure and Applied Chemistry, IUPAC, now recommends that the groups in the periodic table should be numbered 1–18 (Figure 4.1).
- Groups 1 and 2 remain the same as before — classified as the s-block.
- The transition elements now become groups 3–12 — classified as the d-block.
- Groups 3–7 now become groups 13 to 17 and the noble gases become group 18 — classified as the p-block.

The repeating pattern across different periods is known as **periodicity**.

Figure 4.1 The periodic table

Exam practice answers and quick quizzes at **www.hoddereducation.co.uk/myrevisionnotes**

Trends in the periodic table

Atomic radius *decreases* across a period because the attraction between the nucleus and outer electrons increases. This is because:

- the nuclear charge increases
- the outer electrons are being added to the same shell, so there is no extra shielding

Atomic radius *increases* down a group because the attraction between the nucleus and outer electrons decreases. This is because:

- extra shells are added, resulting in the outer shell being further from the nucleus
- there are more shells between the outer electrons and the nucleus, hence there is greater shielding

Electrical conductivity, **melting point** and **boiling points** can be related to structure and bonding, as shown in Figure 4.2.

Giant structures				Molecular structures		
Na	Mg	Al	Si	P_4	S_8	Cl_2
Strong forces between atoms				Weak forces between molecules		
Metallic			Covalent	Dipole–dipole interactions (van der Waals)		
High melting points				Low melting points		
Good conductors			Poor conductors			

Figure 4.2

Electrical conductivity is related to bonding. The elements of groups 1, 2 and 3 are metals. They are good conductors because they contain mobile, delocalised electrons. The outer shell electrons contribute to the mobile, delocalised electrons, which allow metals to conduct heat and electricity, even in the solid state.

The elements in the remaining groups across periods 2 and 3 are poor conductors because they do not have any mobile, free electrons. (Graphite and graphene are exceptions to this and are good conductors because they have mobile free electrons.)

Group 2 elements tend to be better conductors than group 1 because they have two outer shell electrons while group 1 elements only have one outer shell electron.

Melting points and **boiling points** (Figure 4.3) show a gradual increase from group 1 to group 14 followed by a sharp drop to groups 15, 16 and 17. This drop signifies the move from giant structures in groups 1, 2, 3 and 14 to simple molecular structures in groups 15 to 17.

> **Exam tip**
>
> You may be asked to explain the difference between either the melting points or boiling points of two substances — for example, SiO_2 (melting point = 2200°C) and $SiCl_4$ (melting point = –70°C). This is straightforward. If a compound has a high melting point (SiO_2), it is usually because it exists as a giant lattice with strong bonds throughout. If a compound has a low melting point ($SiCl_4$), it is usually because it exists as simple molecules with weak intermolecular forces.

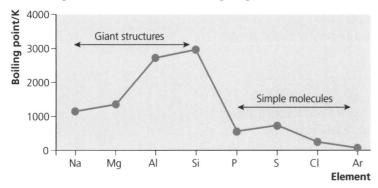

Figure 4.3 Boiling points of the elements across period 3

Ionisation energy *decreases* down a group because:
- Factor 1 — the outer electrons are further from the nucleus, therefore it is easier to remove an electron.
- Factor 2 — shielding increases because of additional inner shells, therefore it is easier to remove an electron.
- Factor 3 — there are more protons in the nucleus, making it harder to remove an electron.

This results in a decrease in the effective nuclear charge because factors 1 and 2 outweigh factor 3.

Ionisation energy *increases* across a period because:
- Factor 1 — the outer electrons are closer to the nucleus, therefore it is harder to remove an electron.
- Factor 2 — shielding is by the same number of inner shells, so it has no effect.
- Factor 3 — there are more protons in the nucleus, making it harder to remove an electron.

This results in an increase in the effective nuclear charge.

Now test yourself

TESTED

1 For each of the following pairs of elements, state which element has the higher first ionisation energy and explain your answer.
  (a) Mg and Na
  (b) Mg and Ca
  (c) Ne and Na
2 The first seven successive ionisation energies of an element, M, are shown in Table 4.1.

Table 4.1

	1	2	3	4	5	6	7
Successive IE/kJ mol⁻¹	790	1600	3200	4400	16 100	19 800	23 800

Suggest in which group of the periodic table you would expect to find element M. Explain your reasoning.

Answers on p. 115

Group 2

Redox reactions of group 2 metals

REVISED

Electron configuration

Each group 2 element has two electrons in its outer shell and readily forms a 2+ ion that has the same electron configuration as a noble gas. It follows that group 2 elements are oxidised when they react. You should be able to use **oxidation number** to illustrate the redox reactions that occur when group 2 elements react with oxygen and with water. Redox is covered on pages 34–36.

Physical properties

Group 2 elements are metals and are, therefore, good conductors. They have reasonably high melting and boiling points. They generally form ionic compounds that are good conductors when molten or aqueous, but poor conductors when solid.

Reaction with O_2

Magnesium, calcium, strontium and barium all react with oxygen to produce an oxide. Reactivity increases down the group, which is because of the increasing ease in which the group 2 element forms the corresponding 2+ ion.

$Mg(s) + \frac{1}{2}O_2(g) \rightarrow MgO(s)$	Burns with a bright white light
$Ca(s) + \frac{1}{2}O_2(g) \rightarrow CaO(s)$	Burns with a brick red colour
$Sr(s) + \frac{1}{2}O_2(g) \rightarrow SrO(s)$	Burns with a crimson colour
$Ba(s) + \frac{1}{2}O_2(g) \rightarrow BaO(s)$	Burns with a light green colour

Each of the above reactions is a redox reaction in which the oxidation number of the group 2 element increases from 0 to +2 and the oxidation number of oxygen decreases from 0 to −2.

Reaction with water

Group 2 elements also undergo redox reactions with water. The oxidation number of the group 2 element increases from 0 to +2 and the oxidation number of hydrogen decreases from +1 to 0:

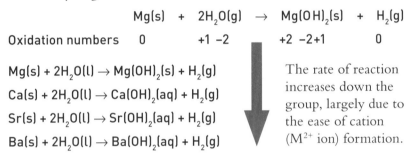

	$Mg(s)$	$+$	$2H_2O(g)$	$\rightarrow$	$Mg(OH)_2(s)$	$+$	$H_2(g)$
Oxidation numbers	0		+1 −2		+2 −2 +1		0

$Mg(s) + 2H_2O(l) \rightarrow Mg(OH)_2(s) + H_2(g)$
$Ca(s) + 2H_2O(l) \rightarrow Ca(OH)_2(aq) + H_2(g)$
$Sr(s) + 2H_2O(l) \rightarrow Sr(OH)_2(aq) + H_2(g)$
$Ba(s) + 2H_2O(l) \rightarrow Ba(OH)_2(aq) + H_2(g)$

The rate of reaction increases down the group, largely due to the ease of cation (M^{2+} ion) formation.

The reaction between magnesium and water is slow. The resultant $Mg(OH)_2$ is barely soluble in water and forms a white suspension.

Magnesium reacts with steam to produce magnesium oxide and hydrogen:

$Mg(s) + H_2O(g) \rightarrow MgO(s) + H_2(g)$

Reaction with dilute acids

Group 2 elements react readily with dilute acids to form a salt and hydrogen. The reactivity again increases down the group. The equation for the reaction varies depending on the acid:

$Mg(s) + 2HCl(aq) \rightarrow MgCl_2(aq) + H_2(g)$

$Mg(s) + H_2SO_4(aq) \rightarrow MgSO_4(aq) + H_2(g)$

But the ionic equation is the same irrespective of the acid:

$Mg(s) + 2H^+(aq) \rightarrow Mg^{2+}(aq) + H_2(g)$

Now test yourself

3 With the aid of equations, identify two redox reactions of calcium. State what has been oxidised in each.

4 Group 2 metals react with aqueous acids to form a salt and hydrogen. Give the formula of the salt formed when:
  (a) barium reacts with nitric acid
  (b) strontium react with ethanoic acid
  (c) calcium reacts with phosphoric acid, H_3PO_4

Answers on p. 115

Reactions of group 2 compounds

REVISED

Reaction of the oxides with water

All group 2 metal oxides react with water to form hydroxides:

$MgO(s) + H_2O(l) \rightarrow Mg(OH)_2(s)$ A suspension is formed

$CaO(s) + H_2O(l) \rightarrow Ca(OH)_2(aq)$ $Ca(OH)_2(aq)$ is known as limewater

$SrO(s) + H_2O(l) \rightarrow Sr(OH)_2(aq)$

$BaO(s) + H_2O(l) \rightarrow Ba(OH)_2(aq)$

These are *not* redox reactions. The oxidation numbers of all the elements are unaltered, for example:

$$MgO(s) \quad + \quad 2H_2O(g) \quad \rightarrow \quad Mg(OH)_2(s)$$
Oxidation numbers   +2 −2   +1 −2   +2 −2+1

The resulting hydroxide solutions are alkaline and have pH values in the region 8–12. The pH varies depending on the concentration of the solution. Calcium hydroxide is used in agriculture to neutralise acidic soils; magnesium hydroxide is used in some indigestion tablets as an antacid.

> **Exam tip**
>
> A question may ask what you would observe when a certain reaction occurs. Use the state symbols as a guide and remember that you will only see effervescence (bubbles) if a gas is produced.

Thermal decomposition of group 2 carbonates

The carbonates are all decomposed to form oxides and carbon dioxide:

$MgCO_3 \rightarrow MgO + CO_2$ Easy to decompose

$CaCO_3 \rightarrow CaO + CO_2$

$SrCO_3 \rightarrow SrO + CO_2$

$BaCO_3 \rightarrow BaO + CO_2$ Hard to decompose

> **Exam tip**
>
> The easiest way to ensure that decomposition is complete is to heat to constant mass.

These are *not* redox reactions. The oxidation numbers of all the elements are unaltered, for example:

$$MgCO_3(s) \quad \rightarrow \quad MgO(s) + CO_2(g)$$
Oxidation numbers   +2 +4 −2   +2 −2   +4 −2

> **Revision activity**
>
> On a postcard write a summary of the reactions of group 2 metals and their compounds.

Now test yourself

TESTED

5 Write an equation, including state symbols, for the reaction between strontium carbonate and nitric acid.

Answer on p. 115

The halogens (group 17)

The halogens are elements that have seven electrons in their outer shells and are in group 17 (formerly group 7) of the periodic table. The halogens exist as simple diatomic molecules.

Electron configuration

Each group 17 element has seven electrons in its outer shell and readily forms a 1– ion (an anion) that has the same electron configuration as a noble gas.

$_9$F $\quad 1s^2 2s^2 2p^5$

$_{17}$Cl $\quad 1s^2 2s^2 2p^6 3s^2 3p^5$

$_{35}$Br $\quad 1s^2 2s^2 2p^6 3s^2 3p^6 3d^{10} 4s^2 4p^5$

$_{53}$I $\quad 1s^2 2s^2 2p^6 3s^2 3p^6 3d^{10} 4s^2 4p^6 4d^{10} 5s^2 5p^5$

Physical properties

Table 4.2 Physical properties of the group 17 elements

Element	State at room temperature	Colour	Volatility
Fluorine, F_2	Gas	Yellow	Down the group there is an increase in induced dipole–dipole interactions, which corresponds to the increased number of electrons in the halogen molecules
Chlorine, Cl_2	Gas	Green	
Bromine, Br_2	Liquid	Orange/brown	
Iodine, I_2	Solid	Grey/black	This increase reduces the volatility and, therefore, increases the melting and boiling point
All are non-metallic, so are poor conductors			

Redox reactions and trends in reactivity

The reactivity of the halogens decreases down the group. This is opposite to the reactivity of the group 2 elements. Group 2 metals react by losing electrons and on descending the group it becomes easier to lose electrons. Halogens react by gaining electrons to form halide anions. The ease of gaining the electron decreases down group 17. This is because atomic radius and shielding increase down the group, and this reduces the effective nuclear attraction for electrons.

Fluorine is a powerful oxidising agent and readily gains electrons.

$F_2 + 2e^- \rightarrow 2F^-$ **Most reactive**

$Cl_2 + 2e^- \rightarrow 2Cl^-$

$Br_2 + 2e^- \rightarrow 2Br^-$

$I_2 + 2e^- \rightarrow 2I^-$ **Least reactive**

> **Exam tip**
>
> It is important to make sure that you know the difference between a halogen and a halide. In exams, many students confuse chloride with chlorine.

Displacement reactions

A halogen (F_2, Cl_2 and Br_2) can displace a halide ion (Cl^-, Br^- and I^-) from one of its salts, as shown in Table 4.3.

Table 4.3 Displacement reactions

	Fluoride, F^-	Chloride, Cl^-	Bromide, Br^-	Iodide, I^-
Fluorine, F_2		Yes	Yes	Yes
Chlorine, Cl_2	No		Yes	Yes
Bromine, Br_2	No	No		Yes
Iodine, I_2	No	No	No	

Displacement reactions illustrate the decrease in oxidising power down group 7.

Chlorine oxidises both bromide and iodide ions. The ionic equation for the oxidation of bromide is:

$$Cl_2(aq) + 2Br^-(aq) \rightarrow 2Cl^-(aq) + Br_2(aq)$$

During the reaction, the orange-brown colour of bromine appears.

The ionic equation for the oxidation of iodide is:

$$Cl_2(aq) + 2I^-(aq) \rightarrow 2Cl^-(aq) + I_2(aq)$$

During the reaction, the brown–black colour of iodine appears. On adding an organic solvent, the solution turns a distinctive violet-purple colour.

Bromine oxidises I^- only:

$$Br_2(aq) + 2I^-(aq) \rightarrow 2Br^-(aq) + I_2(aq)$$

Iodine does *not* oxidise either chloride or bromide.

Each of the displacement reactions is a redox reaction. In each case, the halogen higher in the group gains electrons (is reduced) to form the corresponding halide ion (Figure 4.4).

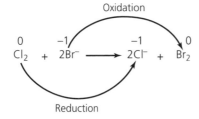

Figure 4.4

Uses of chlorine

Chlorine is used in the treatment of water. Chlorine reacts with water in a reversible reaction and the resultant mixture kills bacteria:

$$Cl_2(aq) + H_2O(l) \rightleftharpoons HCl(aq) + HClO(aq)$$

The reaction is a redox reaction, but it is unusual in that chlorine undergoes both **oxidation** and **reduction** (Figure 4.5).

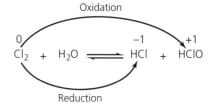

Figure 4.5

One chlorine atom in the Cl_2 molecule is oxidised. Its oxidation number changes from 0 to +1. The other chlorine atom is reduced. Its oxidation number changes from 0 to −1. This type of reaction is called **disproportionation**.

Chlorine is used in water treatment to kill bacteria and make the water safe to drink. This has to be weighed against the possible risks because chlorine is toxic and reacts with substances such as hydrocarbons to form chlorinated hydrocarbons. If drinking water contained hydrocarbons this could present a risk to health.

Chlorine also reacts with sodium hydroxide to form bleach, which is a mixture of sodium chloride and sodium chlorate(I). This is also a disproportionation reaction of chlorine (Figure 4.6):

$$Cl_2(g) + NaOH(aq) \rightarrow NaCl(aq) + NaClO(aq)$$

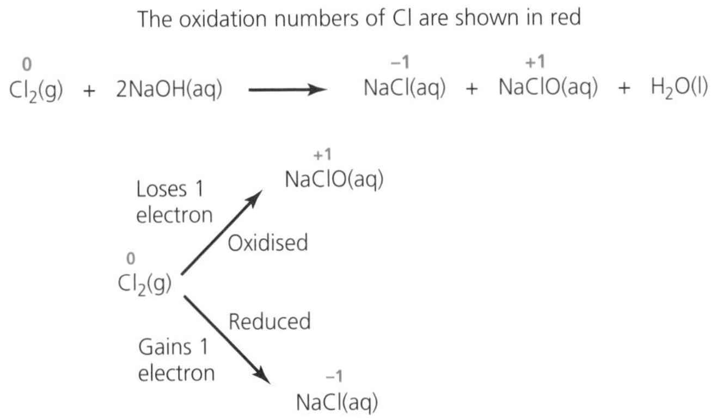

The oxidation numbers of Cl are shown in red

$$\overset{0}{Cl_2}(g) + 2NaOH(aq) \longrightarrow \overset{-1}{Na}Cl(aq) + \overset{+1}{Na}ClO(aq) + H_2O(l)$$

Loses 1 electron — Oxidised → $\overset{+1}{Na}ClO(aq)$

$\overset{0}{Cl_2}(g)$

Gains 1 electron — Reduced → $\overset{-1}{Na}Cl(aq)$

Figure 4.6

Chlorine reacts with NaOH(aq) to form a number of different chlorates, including sodium chlorate(I), NaClO; sodium chlorate(III), $NaClO_2$; and sodium chlorate(v), $NaClO_3$. In each of these reactions NaCl and water are also formed. Each reaction is a disproportionation reaction.

Example

When chlorine reacts with a hot concentrated solution of NaOH(aq), sodium chlorate(v), $NaClO_3$(aq), is formed. Construct an equation for this reaction.

Answer

We know that NaCl(aq) will be formed along with $NaClO_3$(aq). The oxidation changes in Cl are −1 in NaCl and +5 in $NaClO_3$ (Figure 4.7).

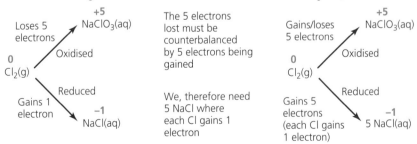

Loses 5 electrons — Oxidised → $\overset{+5}{Na}ClO_3(aq)$

$\overset{0}{Cl_2}(g)$

Gains 1 electron — Reduced → $\overset{-1}{Na}Cl(aq)$

The 5 electrons lost must be counterbalanced by 5 electrons being gained

We, therefore need 5 NaCl where each Cl gains 1 electron

Gains/loses 5 electrons — Oxidised → $\overset{+5}{Na}ClO_3(aq)$

$\overset{0}{Cl_2}(g)$

Gains 5 electrons (each Cl gains 1 electron) — Reduced → $\overset{-1}{5Na}Cl(aq)$

Figure 4.7

We now know that the products will contain $1NaClO_3$(aq), 5NaCl(aq) and H_2O(l). So that the balanced equation can be constructed to give:

$$3Cl_2(g) + 6NaOH(aq) \rightarrow NaClO(aq) + 5NaCl(aq) + 3H_2O(l)$$

It is important to remember that oxidation number changes have to be balanced as well as balancing symbols.

Disproportionation is the simultaneous oxidation and reduction of an element such that during a reaction its oxidation number both increases and decreases.

Now test yourself

TESTED

6 Write an equation for the reaction between chlorine and NaOH to produce $NaClO_2$, NaCl and H_2O.

Answer on p. 115

Reactions of halide ions

REVISED

Silver chloride, silver bromide and silver iodide are insoluble in water. Therefore, the presence of chloride, bromide or iodide ions in a solution can be detected by the addition of a solution of silver nitrate ($AgNO_3(aq)$). Each of the silver halides forms a different coloured precipitate. Each of the precipitates can be distinguished by their solubility in ammonia.

$$Ag^+(aq) + Cl^-(aq) \rightarrow AgCl(s)$$

AgCl is a white precipitate, which is soluble in dilute NH_3.

$$Ag^+(aq) + Br^-(aq) \rightarrow AgBr(s)$$

AgBr is a cream precipitate, which is soluble in concentrated NH_3.

$$Ag^+(aq) + I^-(aq) \rightarrow AgI(s)$$

AgI is a yellow precipitate, which is insoluble in concentrated NH_3.

> **Revision activity**
>
> On a postcard write a summary of the reactions of group 17 the halogens and their compounds.

Now test yourself

TESTED

7 Chlorine reacts explosively with ethyne, C_2H_2, to form carbon and hydrogen chloride.
  (a) Construct an equation, including state symbols, for this reaction.
  (b) Using oxidation numbers, explain the role of chlorine.
  (c) If chlorine is replaced by fluorine, would the reaction be more or less explosive? Explain your answer.

Answer on p. 115

Qualitative analysis

You will be expected to analyse and detect a range of ions by a series of test tube reactions. Table 4.4 details qualitative tests for a range of ions.

Table 4.4

Ion	Test	Equation	Observation
CO_3^{2-}	Add an acid, $H^+(aq)$	$CO_3^{2-}(aq) + 2H^+(aq) \rightarrow CO_2(g) + H_2O(l)$	Effervescence, bubbles
SO_4^{2-}	Add aqueous $BaCl_2(aq)$	$SO_4^{2-}(aq) + Ba^{2+}(aq) \rightarrow BaSO_4(s)$	White precipitate
Cl^-		$Cl^-(aq) + Ag^+(aq) \rightarrow AgCl(s)$	White precipitate*
Br^-	Add $AgNO_3(aq)$	$Br^-(aq) + Ag^+(aq) \rightarrow AgBr(s)$	Cream precipitate*
I^-		$I^-(aq) + Ag^+(aq) \rightarrow AgI(s)$	Yellow precipitate*
NH_4^+	Warm with NaOH(aq)	$NH_4^+ + OH^-(aq) \rightarrow NH_3(g) + H_2O(l)$	$NH_3(g)$ is evolved which turns moist red litmus blue

*The silver halide precipitates can be further distinguished by adding ammonia to the precipitates. See above.

Exam practice answers and quick quizzes at **www.hoddereducation.co.uk/myrevisionnotes**

Now test yourself

8 Barium carbonate, barium chloride and barium sulfate are all white solids. Suggest a series of reactions that could be used to distinguish which is which. Write ionic equations for each reaction.

Answer on p. 115

Revision activity

On a postcard write a summary of the observations that you would see when testing for anions and cations.

Exam practice

1 (a) Chlorine bleach is made by the reaction of chlorine with aqueous sodium hydroxide. In this reaction the oxidation number of Cl_2 changes and Cl_2 is said to undergo disproportionation.

$$Cl_2(g) + 2NaOH(aq) \rightarrow NaClO(aq) + NaCl(aq) + H_2O(l)$$

   (i) Determine the oxidation number of chlorine in Cl_2, NaClO and NaCl. [1]

   (ii) State what is meant by the term *disproportionation*. [1]

   (iii) The bleaching agent is the ClO^- ion. In the presence of sunlight, this ion decomposes to release oxygen gas. Construct an equation for this reaction. [2]

(b) The sea contains a low concentration of bromide ions. Bromine can be extracted from seawater by first concentrating the seawater and then bubbling chlorine through this solution.

   (i) Suggest how the seawater could be concentrated [1]

   (ii) The chlorine oxidises bromide ions to bromine. Construct a balanced ionic equation for this reaction. [1]

(c) Vinyl chloride is a compound of chlorine, carbon and hydrogen. It is used to make polyvinylchloride (PVC). Vinyl chloride has the percentage composition by mass: chlorine, 56.8%; carbon, 38.4 %; hydrogen, 4.8%.

   (i) Show that the empirical formula of vinyl chloride is ClC_2H_3. Show your working. [2]

   (ii) The molecular formula of vinyl chloride is the same as its empirical formula. Draw a possible structure, including bond angles, for a molecule of vinyl chloride. [2]

2 (a) Barium is a group 2 element. It reacts with oxygen to form compound **A** and with water to form compound **B** and gas **C**. With the aid of suitable equations, identify, **A**, **B** and **C**. [3]

(b) (i) Write an equation, including state symbols, for the thermal decomposition of barium carbonate. [2]

   (ii) Calculate the minimum volume of $0.05\,mol\,dm^{-3}$ HCl(aq) that is needed to react with 1.00 g of barium carbonate. Show all your working. [4]

   (iii) Explain why *more* $0.05\,mol\,dm^{-3}$ HCl(aq) would be needed to react with 1.00 g of magnesium carbonate than with 1.00 g of barium carbonate. Show all your working. [2]

Answers and quick quiz 4 online

ONLINE ☐

Summary

You should now have an understanding of:
- trends in atomic radii
- trends in melting points and boiling points
- trends in ionisation energies
- reactions of group 2 elements with oxygen and with water
- reactions of group 2 oxides with water
- decomposition of group 2 carbonates

- physical properties of group 17 elements
- displacement reactions of the halogens and the halides
- reactions of halides with Ag^+ and NH_3
- disproportionation reactions of chlorine
- qualitative analysis for carbonate, sulfate, halides and ammonium ions

5 Physical chemistry

Enthalpy changes

ΔH of reaction, formation, combustion and neutralisation

Enthalpy change, ΔH, is the exchange of energy between a reaction mixture and its surroundings. It is measured at constant temperature and constant pressure. The units are $kJ\,mol^{-1}$.

ΔH can be calculated using the equation:

ΔH = enthalpy of products − enthalpy of reactants

Enthalpy changes can be represented by simple enthalpy profile diagrams.

For an exothermic reaction, the enthalpy profile diagram shows the products at a lower energy than the reactants. For an endothermic reaction the enthalpy profile diagram shows the products at a higher energy than the reactants. The difference in the enthalpy is ΔH. E_a is the **activation energy** (Figure 5.1).

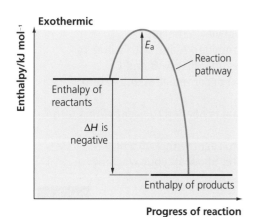
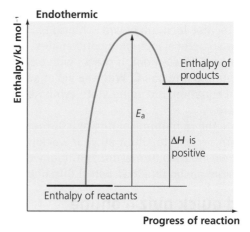

Figure 5.1 Enthalpy profiles for an exothermic and an endothermic reaction

Oxidation reactions such as the combustion of fuels are exothermic and release energy to the surroundings. This results in an increase in temperature of the surroundings. The enthalpy profile in Figure 5.2 illustrates the combustion of methane.

Activation energy is the *minimum* amount of energy required to start the reaction.

Standard state is the physical state of a substance at 298 K and 101 kPa.

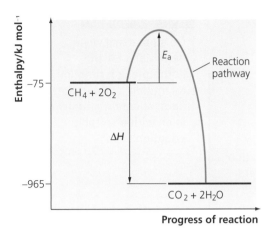

Figure 5.2 Enthalpy profile for the combustion of methane

The activation energy is the minimum energy needed for colliding particles to react. In any chemical reaction bonds are broken and new bonds are formed. Breaking bonds is an endothermic process that requires energy. This energy requirement contributes to the activation energy of a reaction.

Standard enthalpy changes

All **standard enthalpy changes** relate to substances in their **standard states** and are, therefore, measured under standard conditions. The standard temperature is 298 K (25°C) and the standard pressure is 101 kPa (for exam purposes 100 kPa is acceptable). Standard temperature and pressure are abbreviated to s.t.p.

Examinations often ask for a definition of an enthalpy change. It is advisable to learn the definitions on this page.

You may also have to show your understanding by writing equations that illustrate the standard enthalpy changes of reaction, formation and combustion.

Rules for writing equations to illustrate standard enthalpies of formation and combustion are shown in Table 5.1.

Table 5.1

Standard enthalpy of formation	Standard enthalpy of combustion
Show the elements as the reactants	React **1** mol of the substance with excess oxygen, even if this means having fractions in the equation
Produce **1** mol of the substance, even if this means having fractions in the equation	
	Show the state symbols
Show the state symbols	
Example:	Example:
$2C(s) + 3H_2(g) + \frac{1}{2}O_2(g) \rightarrow$ $CH_3CH_2OH(l)$	$C_2H_6(g) + 3\frac{1}{2}O_2 \rightarrow 2CO_2(g) + 3H_2O(l)$

The **standard enthalpy of neutralisation** is the enthalpy change when 1 mol of water is formed from the reaction of an acid and base under standard conditions of 298 K and 101 kPa. The equation for the standard enthalpy of neutralisation of dilute hydrochloric acid by sodium hydroxide solution is:

$$HCl(aq) + NaOH(aq) \rightarrow NaCl(aq) + H_2O(l)$$

Standard enthalpy change of reaction, $\Delta_r H^\ominus$, is the enthalpy change when the number of moles of the substances in the balanced equation react under standard conditions of 298 K and 101 kPa.

Standard enthalpy change of formation $\Delta_f H^\ominus$, is the enthalpy change when 1 mol of a substance is formed from its elements, in their natural state, under standard conditions of 298 K and 101 kPa.

Standard enthalpy change of combustion, $\Delta_c H^\ominus$, is the enthalpy change when 1 mol of a substance is burnt completely, in an excess of oxygen, under standard conditions of 298 K and 101 kPa.

Standard enthalpy change of neutralisation, $\Delta_{neut} H^\ominus$, is the enthalpy change when 1 mol of water is formed in a neutralisation reaction (a reaction between an acid and a base) under standard conditions of 298 K and 101 kPa.

Standard state is the physical state of a substance at 298 K and 101 kPa.

An ionic equation can also be used to show the standard enthalpy of neutralisation:

$$H^+(aq) + OH^-(aq) \rightarrow H_2O(l)$$

Typical mistake

If you are asked to write an equation to illustrate the enthalpy of neutralisation using sulfuric acid and sodium hydroxide, do *not* write: $H_2SO_4(aq) + 2NaOH(aq) \rightarrow Na_2SO_4(aq) + 2H_2O(l)$, because this equation shows the formation of two moles of water and by definition the enthalpy of neutralisation forms one mole of water. The correct equation is:

$$\frac{1}{2}H_2SO_4(aq) + NaOH(aq) \rightarrow \frac{1}{2}Na_2SO_4(aq) + H_2O(l)$$

Revision activity

On a postcard write definitions for the enthalpies of: reaction, formation, combustion and neutralisation. Illustrate each with a suitable equation.

Now test yourself

TESTED

1 Define standard enthalpy change of formation. Write an equation to illustrate the standard enthalpy change of formation of propanal, $CH_3CH_2CHO(l)$.
2 Define standard enthalpy change of combustion. Write an equation to illustrate the standard enthalpy change of combustion of propanone, $CH_3COCH_3(l)$.

Answer on p. 116

Enthalpy changes using experimental data

REVISED

The standard enthalpy change, $\Delta_r H^\ominus$, for reactions that take place in solution can usually be measured directly using the simple apparatus shown in Figure 5.3. The result obtained is only approximate because there will be heat losses to the surroundings.

The energy transfer for the reaction mixture is given the symbol q and can be calculated using the equation:

$$q = mc\Delta T$$

where m is the mass of the reaction mixture, c is the specific heat capacity of the reaction mixture and ΔT is the change in temperature.

The enthalpy change for the reaction mixture, q, has a value in either J (joules) or kJ (kilojoules) depending on the units of specific heat capacity. It is usual to adjust this value so that ΔH can be quoted for 1 mol of reactant with the units of kJ mol^{-1}. The standard enthalpy change, ΔH_r, for the reaction is calculated by dividing the energy transferred by the number of moles, n, of reactant used:

$$\Delta_r H^\ominus = \frac{q}{n} = \frac{mc\Delta T}{n}$$

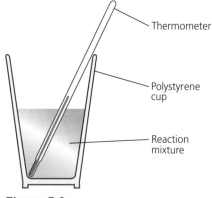

Thermometer

Polystyrene cup

Reaction mixture

Figure 5.3

Example

When 50.0 cm³ of 2.00 mol dm⁻³ hydrochloric acid is mixed with 50.0 cm³ of 2.00 mol dm⁻³ sodium hydroxide solution, the temperature increased by 13.7°C.

Calculate the standard enthalpy change of neutralisation of hydrochloric acid. Assume that the specific heat capacity, c, = 4.20 J g⁻¹K⁻¹ and that the density of both the hydrochloric acid and sodium hydroxide solution is both 1.00 g cm⁻³.

Answer

Step 1: calculate the energy transferred in the reaction by using:

$q = mc\Delta T$

m = mass of the two solutions = 50.0 + 50.0 = 100.0 g

$q = mc\Delta T = 100.0 \times 4.20 \times 13.7 = 5754\,J = 5.754\,kJ$

Step 2: convert the answer to kJ mol^{-1} by dividing by the number of moles used:

amount, n, in moles of HCl = $cV = 2.00 \times \dfrac{50}{1000} = 0.1$ mol

$\Delta_{neut}H = \dfrac{q}{n} = \dfrac{5.754}{0.1} = 57.54 = = 57.5\,kJ\,mol^{-1}$

Remember that because the temperature went up it is an exothermic reaction. Therefore:

$\Delta_{neut}H = -57.5\,kJ\,mol^{-1}$

> **Typical mistake**
>
> In a question such as '2.30 g ethanol was burnt and raised the temperature of 300 cm³ of water by 12.5°C. Calculate $\Delta_c H^\ominus$ for ethanol', the density of water and the specific heat capacity, c, of the apparatus would be given. In the first step, when using $q = mc\Delta T$, many students make the mistake of using 2.30 g as the mass. The mass used should be the mass of water.

Now test yourself

TESTED

3 When 25.0 cm³ of 1.00 mol dm^{-3} nitric acid was mixed with 50.0 cm³ of 0.50 mol dm^{-3} sodium hydroxide solution the temperature increased by 4.6°C. Calculate the enthalpy change of neutralisation of nitric acid. Assume that the specific heat capacity, c = 4.20 J g^{-1} K^{-1} and that the density of both nitric acid and sodium hydroxide solution is 1.00 g cm^{-3}.

Answer on p. 116

Enthalpy changes using average bond enthalpy data

REVISED

Breaking a bond requires energy; forming a bond releases energy. The energy required to break a bond has the same numerical value as the energy released when the bond is formed.

Bond enthalpies are the *average* (mean) values and do *not* take into account the specific chemical environment. Some average bond enthalpies are shown in Table 5.2.

Table 5.2 Some average bond enthalpies

Bond	C–H	C=O	O=O	O–H	C–N	C=C	C–C	H–H
ΔH/kJ mol^{-1}	+413	+805	+498	+464	+286	+612	+347	+436

> **Bond enthalpy** is the enthalpy change required to break the same bond in all the molecules in 1 mol of a gas and to separate the resulting gaseous (neutral) particles/atoms/radicals so they exert no forces on each other. It is best reinforced by a simple equation such as:
>
> $Cl–Cl(g) \rightarrow Cl\bullet(g) + Cl\bullet(g)$

Calculations involving average bond enthalpy

REVISED

Calculations are straightforward. In order for a reaction to take place existing bonds have first to be broken (energy is required, so endothermic) and then new bonds have to be formed (energy is given out, so exothermic).

The enthalpy of combustion of methane can be calculated using average bond enthalpies. The equation is:

$CH_4 + 2O_2 \rightarrow CO_2 + 2H_2O$

It is useful to draw out the reaction using displayed formulae so that all the bonds broken and formed can be seen clearly (Figure 5.4).

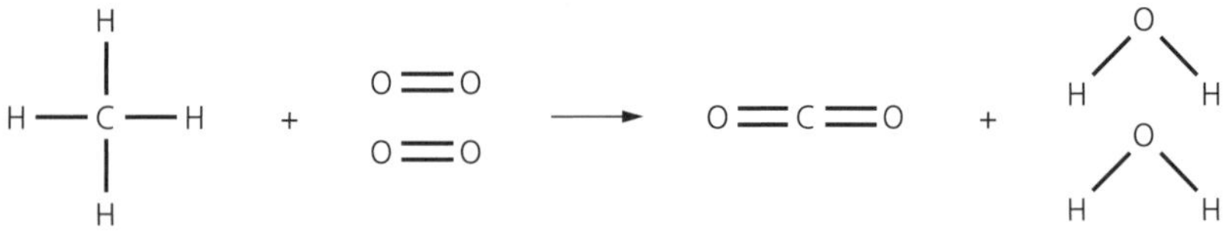

Figure 5.4

Bonds broken:

$4 \times (C–H) = 4 \times +413 = +1652\,kJ\,mol^{-1}$

$2 \times (O=O) = 2 \times +498 = +996\,kJ\,mol^{-1}$

total energy needed to *break* all bonds = +2648 kJ mol⁻¹

Bonds formed:

$2 \times (C=O) = 2 \times -805 = -1610\,kJ\,mol^{-1}$

$4 \times (O–H) = 4 \times -464 = -1856\,kJ\,mol^{-1}$

total energy released to *form* all bonds = –3466 kJ mol⁻¹

Enthalpy change for the reaction:

$\Delta H = +2648 - 3466 = -818\,kJ\,mol^{-1}$

The accepted value for this reaction is $-890\,kJ\,mol^{-1}$, which differs substantially from $-818\,kJ\,mol^{-1}$. This is largely explained by using average bond enthalpies for the C–H, C=O and O–H bonds in the calculation and not specific bond enthalpies.

Enthalpy changes using Hess's law

REVISED

For energetic (the activation energy is very high) or kinetic (the reaction rate is very slow) reasons, the enthalpy changes for many chemical reactions cannot be measured directly by experiment, but they can be calculated using **Hess's law**.

The enthalpy of formation of $CO(g)$ cannot be measured directly but it can be calculated using Hess's law.

The enthalpy change for the following reactions can be measured experimentally:

$C(s) + O_2(g) \rightarrow CO_2(g)$ $\Delta H = -394\,kJ\,mol^{-1}$

$CO(g) + \frac{1}{2}O_2(g) \rightarrow CO_2(g)$ $\Delta H = -284\,kJ\,mol^{-1}$

> **Hess's law** states that if a reaction can take place by more than one route, the overall enthalpy change for the reaction is the same irrespective of the route taken, provided that the initial and final conditions are the same.

When applying Hess's law it is helpful to construct an enthalpy triangle.

Step 1: start with the enthalpy change that has to be calculated. Call it ΔH_1. Write an equation for the reaction. This is the top line of the triangle.

$C(s) + \frac{1}{2}O_2(g) \xrightarrow{\Delta H_1} CO(g)$

Step 2: construct a cycle with two alternative routes:

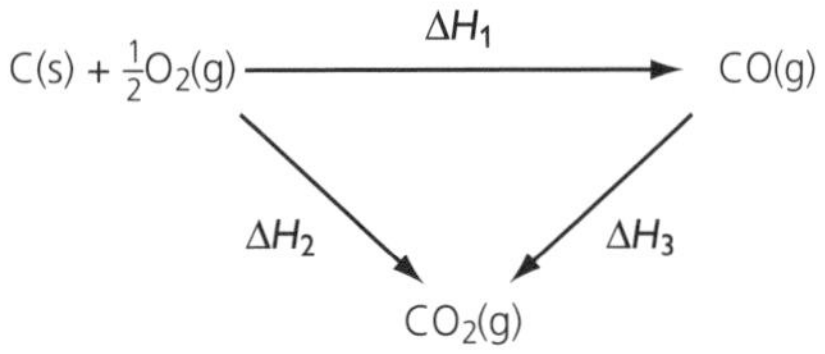

Exam practice answers and quick quizzes at **www.hoddereducation.co.uk/myrevisionnotes**

Step 3: apply Hess's law to the triangle:

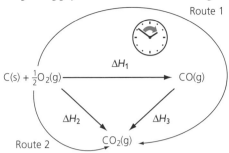

Look at the direction of the arrows. Route 1 has arrows that point in the clockwise direction; route 2 has arrows that point in the anti-clockwise direction.

route 1 = route 2

$\Delta H_1 + \Delta H_3 = \Delta H_2$

$\Delta H_1 = \Delta H_2 - \Delta H_3$

      $= -394 - (-284) = -394 + 284 = -110\,\text{kJ}\,\text{mol}^{-1}$

So, for $C(s) + \frac{1}{2}O_2(g) \rightarrow CO(g)$, $\Delta H = -110\,\text{kJ}\,\text{mol}^{-1}$

The enthalpy of formation of carbon monoxide is $-110\,\text{kJ}\,\text{mol}^{-1}$.

Enthalpy of formation from enthalpies of combustion

If you are asked to calculate the enthalpy change of formation using enthalpies of combustion, it is best to construct a cycle with the combustion products at the bottom. The arrows always point downwards to the combustion products (Figure 5.5).

If Hess's law is applied to the cycle:

    $\Delta H_1 + \Delta H_3 = \Delta H_2$ hence $\Delta H_1 = \Delta H_2 - \Delta H_3$

Figure 5.5

Enthalpy of combustion from enthalpies of formation

If you are asked to calculate the enthalpy change of combustion from enthalpy of formation data, it is best to construct a cycle with the elements at the bottom. The arrows always point upwards (Figure 5.6).

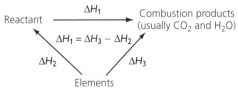

Figure 5.6

If Hess's law is applied to the cycle:

    $\Delta H_1 + \Delta H_2 = \Delta H_3$ hence $\Delta H_1 = \Delta H_3 - \Delta H_2$

Enthalpy of reaction from enthalpies of formation

If you are asked to calculate the enthalpy change of reaction from enthalpy of formation data, it is best to construct a cycle with the elements at the bottom. The arrows always point upwards.

Example

Use the enthalpy of formation data to calculate the enthalpy of reaction for:

$$CH_3COCl(l) + C_2H_5OH(l) \rightarrow CH_3COOC_2H_5(l) + HCl(g)$$

	$CH_3COCl(l)$	$C_2H_5OH(l)$	$CH_3COOC_2H_5(l)$	$HCl(g)$
$\Delta H_f/\text{kJ mol}^{-1}$	−272.9	−277.1	−479.3	−92.3

Answer

Construct a simple triangle with the 'elements' at the bottom (Figure 5.7).

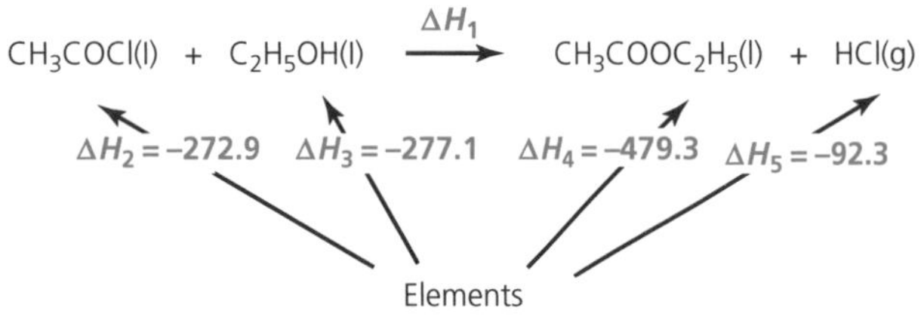

$$CH_3COCl(l) + C_2H_5OH(l) \xrightarrow{\Delta H_1} CH_3COOC_2H_5(l) + HCl(g)$$

$\Delta H_2 = -272.9 \quad \Delta H_3 = -277.1 \quad \Delta H_4 = -479.3 \quad \Delta H_5 = -92.3$

Elements

Figure 5.7

Clockwise steps will balance anti-clockwise steps such that:

$\Delta H_1 + \Delta H_2 + \Delta H_3 = \Delta H_4 + \Delta H_5$ **which can be rearranged to give**

$\Delta H_1 = \Delta H_4 + \Delta H_5 - \Delta H_2 - \Delta H_3 = -479.3 - 92.3 + 272.9 + 277.1 = -21.6\,\text{kJ mol}^{-1}$

Now test yourself

TESTED ☐

4 Use the data in the table below to calculate the standard enthalpy of formation for propane.

Substance	$\Delta H_c^{\ominus}/\text{kJ mol}^{-1}$
C(s)	−394
$H_2(g)$	−286
$C_3H_8(g)$	−2219

Answer on p. 116

> **Exam tip**
>
> Most exam papers contain at least one question using Hess's law. The data provided will have to be either enthalpies of combustion or enthalpies of formation. If the data provided are:
> - standard enthalpies of combustion, the cycle will have combustion products at the bottom and the arrows will point down
> - standard enthalpies of formation, the cycle will have elements at the bottom and the arrows will point up

Reaction rates

Experimental observations show that the rate of a reaction is influenced by temperature, concentration and the use of a catalyst.

Simple collision theory

REVISED ☐

The collision theory of reactivity helps to provide explanations for these observations. A reaction cannot take place unless a collision occurs between reacting particles. Increasing temperature or concentration increases the chance of a collision occurring.

However, not all collisions lead to a successful reaction. The energy of a collision between reacting particles must exceed the minimum energy required for the reaction to occur. This minimum energy is known as the activation energy, E_a. Increasing the temperature affects the number of collisions with energy that exceeds the activation energy. A catalyst effectively lowers the activation energy.

Calculation of reaction rates

Take for example the reaction between calcium carbonate and hydrochloric acid:

$$CaCO_3(s) + 2HCl(aq) \rightarrow CaCl_2(aq) + CO_2(g) + H_2O(l)$$

In an experiment, the volume of carbon dioxide being produced could be measured as the reaction proceeds in a syringe, as shown in Figure 5.8.

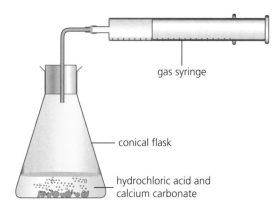

gas syringe

conical flask

hydrochloric acid and calcium carbonate

Figure 5.8 Collection of gas using a gas syringe

Carbon dioxide would initially be produced quickly but, as the reaction proceeded, it would gradually slow down as the reactants were used up. Eventually, once one or both of the reactants had been used up, the production of carbon dioxide would cease. A graph of the volume of carbon dioxide produced against time would appear as shown in Figure 5.9.

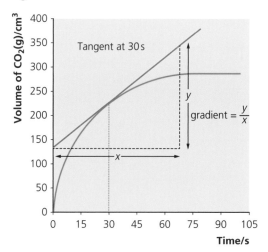

Figure 5.9 Volume of carbon dioxide collected against time

The rate is not constant but can be given a numerical value at any particular time by drawing a tangent to the curve and measuring its gradient. For example, in Figure 5.9 the rate at which the reaction is proceeding at 30 s is given by the gradient of the tangent at that point on the curve. In this case its value is $(350 - 130) \, cm^3/70 \, s = 1.7 \, cm^3 s^{-1}$.

Boltzmann's distribution of molecular energies

Energy is directly proportional to absolute temperature. When collisions occur, the particles involved in the collision exchange (gain or lose) energy, even if a reaction does not occur. It follows that for any given mass of gaseous reactants at constant temperature; there will be a distribution of energies with some particles having more energy than others. Figure 5.10 shows a typical distribution of energies at constant temperature.

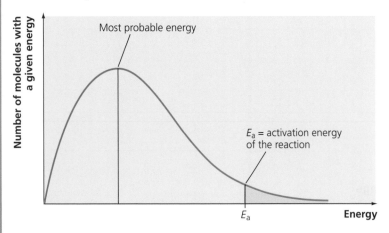

Figure 5.10

- The distribution always goes through the origin.
- The distribution is asymptotic to the horizontal axis at high energy, showing that there is no maximum energy. (Asymptotic means that the curve approaches the axis but will only meet it at infinity.)
- E_a represents the activation energy — the minimum energy required to start the reaction.
- The area under the curve represents the total number of particles.
- The yellow shaded area represents the number of particles with sufficient energy to react. These are particles with energy greater than or equal to the activation energy, $E \geq E_a$ (Figure 5.11).

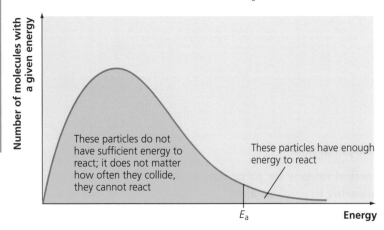

Figure 5.11

Now test yourself

5 Define the term activation energy, E_a.

Answer on p. 116

Effects of concentration, temperature and using a catalyst on the rate of reaction

REVISED

Effect of concentration

A useful analogy is to imagine your first driving lesson. The one thing you want to avoid is a collision! It follows that your first lesson is likely to be early on a Sunday morning on a quiet country lane, rather than at 5.00 p.m. on a Friday evening in the city centre. It is obvious that the high concentration of cars at rush hour increases the chance of a collision. The same is true for a chemical reaction. Increasing concentration simply increases the chance of a collision. The more collisions there are, the faster the reaction will be.

For a gaseous reaction, increasing pressure has the same effect as increasing concentration. When gases react they react faster at high pressure because as the pressure increases so does the concentration and hence there is an increased chance of a collision.

Now test yourself

TESTED

6 Explain how increasing the pressure on a gaseous reaction affects the rate of reaction.

Answer on p. 116

Effect of temperature

An increase in temperature (to T_2 in Figure 5.12) has a dramatic effect on the distribution of energies.

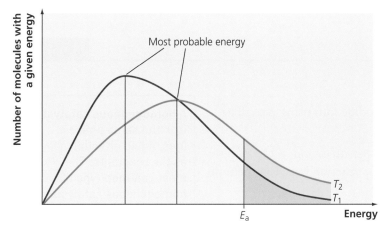

Figure 5.12

Only the temperature has changed. The number of particles is constant, so the area under both curves remains the same.

At higher temperature the distribution flattens and shifts to the right. This shows that increasing temperature increases the number of particles with energy greater than or equal to the activation energy, $E \geq E_a$, which means that at high temperature there are more particles with sufficient energy to react. Therefore, the reaction is faster. Decreasing the temperature has the opposite effect.

> **Exam tip**
>
> For a Boltzmann distribution curve at a higher temperature the marking points are for the following:
> - The curve goes through the origin and there are fewer particles with low energy.
> - The most probable energy moves to right (higher energy), but the height of the peak is lower.
> - There are more particles with high energy, so a greater proportion of particles have energy that exceed the activation energy.

Effect of a catalyst

Catalysts speed up reactions without themselves being changed permanently. Catalysts work by providing an alternative route for the reaction, which has a lower activation energy. This is illustrated in Figure 5.13.

E_a is the activation energy of the uncatalysed reaction, E_{cat} is the activation energy of the catalysed reaction. E_{cat} is lower than E_a.

A catalyst lowers the activation energy but does not alter the Boltzmann distribution (Figure 5.14). It increases the number of particles with energy greater than or equal to the new activation energy, E_{cat}.

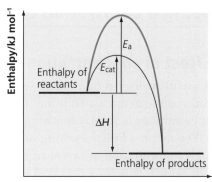

Figure 5.13 Effect of a catalyst on activation energy

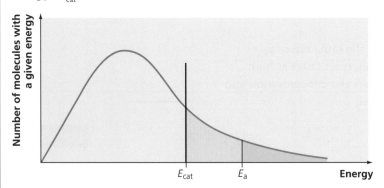

Figure 5.14 Effect of a catalyst on Boltzmann distribution

Now test yourself

TESTED

7 Draw an enthalpy profile diagram and a Boltzmann distribution to show how a catalyst works.

Answer on p. 116

Types of catalyst

REVISED

Homogeneous catalysts

A suitable example involves the oxidation of iodide ions using persulfate ions (peroxosulfate ions).

Iodide ions, $I^-(aq)$, are easily oxidised and the persulfate ion, $S_2O_8^{2-}(aq)$, is a powerful oxidising agent, but surprisingly the reaction between them in water is slow. The equation for the reaction is:

$$S_2O_8^{2-} + 2I^- \rightarrow 2SO_4^{2-} + I_2$$

For the reaction to occur the two ions must collide, however, the rate of collision is slow because both ions are negative and repel each other.

The reaction can be catalysed by adding aqueous solutions of either $Fe^{2+}(aq)$ ions or $Fe^{3+}(aq)$ ions.

The transition metal ions catalyse the reaction by providing an alternate mechanism (Table 5.3).

> A **homogeneous catalyst** is in the same phase (gas, liquid or solid) as the reactants. The most common type of homogeneous catalysis involves reactions in aqueous solutions.

Table 5.3

Catalyst	Fe²⁺(aq)	Fe³⁺(aq)
Step 1	$S_2O_8^{2-} + 2Fe^{2+} \rightarrow 2SO_4^{2-} + 2Fe^{3+}$	$2Fe^{3+} + 2I^- \rightarrow 2Fe^{2+} + I_2$
Step 2	$2Fe^{3+} + 2I^- \rightarrow 2Fe^{2+} + I_2$	$S_2O_8^{2-} + 2Fe^{2+} \rightarrow 2SO_4^{2-} + 2Fe^{3+}$
	Fe²⁺ is reformed and acts as a catalyst	Fe³⁺ is reformed and acts as a catalyst
Net reaction	$S_2O_8^{2-} + 2I^- \rightarrow 2SO_4^{2-} + I_2$	$S_2O_8^{2-} + 2I^- \rightarrow 2SO_4^{2-} + I_2$

The Fe^{2+} and the Fe^{3+} ions provide an alternative mechanism that involves an intermediate step of lower activation energy. The activation energy is lowered because both steps in either mechanism involve ions of opposite charge. The iron ions from the catalyst take part in the reaction, but are released at the end of the reaction, so the final amount of the catalyst is the same as when the reagents were mixed.

Another example of a homogeneous catalyst occurs in the loss of ozone from the upper atmosphere (stratosphere). This is a gas phase reaction in which chlorine radicals act catalytically.

Heterogeneous catalysts

A heterogeneous catalyst works by adsorbing the gases onto its solid surface. This adsorption weakens bonds within the reactant molecules, which lowers the activation energy of the reaction. Bonds are broken and new bonds are formed. The product molecules are desorbed from the solid surface of the catalyst.

Iron is used as the catalyst in the Haber process for the production of ammonia (Figure 5.15).

$$N_2(g) \ + \ 3H_2(g) \ \xrightarrow[\text{as catalyst}]{\text{Fe(s)}} \ 2NH_3(g)$$

Figure 5.15

> A **heterogeneous catalyst** is in a different phase from the reactants. The most common type of heterogeneous catalysis involves reactions of gases in the presence of a solid catalyst.

The internal combustion engine discharges pollutants into the atmosphere. Modern cars are fitted with catalytic converters (made from platinum, rhodium and palladium) that reduce the emission of unburnt hydrocarbons, carbon monoxide and oxides of nitrogen.

Removal of unburnt hydrocarbons:

$$C_8H_{18}(g) + 12\tfrac{1}{2}O_2(g) \rightarrow 8CO_2(g) + 9H_2O(g)$$

Removal of carbon monoxide and nitrogen monoxide:

$$2NO(g) + 2CO(g) \rightarrow N_2(g) + 2CO_2(g)$$

Removal of oxides of nitrogen:

$$2NO_2(g) + 4CO(g) \rightarrow N_2(g) + 4CO_2(g)$$

All of the above reactions are in the gas phase. The catalytic converter is a solid mounted on a fine aluminium mesh.

> **Exam tip**
>
> If you are asked how a heterogeneous catalyst works, look at the marks allocation. If there is more than 1 mark, the explanation required is usually:
> ● The catalyst works by adsorbing gases onto its surface.
> ● This weakens the bonds, which lowers the activation energy and leads to a chemical reaction.
> ● The products of the reaction are desorbed from the surface of the catalyst.

Now test yourself

TESTED ☐

8 Explain what is meant by **heterogeneous** catalysis and by **homogeneous** catalysis. Give an example of each type.

Answer on p. 116

Economic importance of catalysts

Catalysts are of great economic importance. They often allow reactions to be carried out at lower temperatures and/or pressures, which reduce costs. They also enable the use of different reactions with lower atom economies. This also reduces waste. Enzymes are being used increasingly to generate specific products and to enable reactions to be carried out close to room temperature and pressure.

Chemical equilibrium

Reversible reactions

REVISED

There are many everyday examples of reversible reactions or processes, the most common being the changes between the physical states of water. If the temperature of water falls below 0°C, the water freezes and ice forms. When the temperature rises above 0°C, the ice melts and water forms again. This process can be represented as:

$$H_2O(s) \rightleftharpoons H_2O(l)$$

The $\rightleftharpoons$ symbol indicates that a reaction is reversible.

Esterification is an example of a reversible chemical reaction. Esterification is covered in full in Module 6 of the A level course.

Dynamic equilibrium

In a reversible reaction (Figure 5.15), the reaction from left to right is called the forward reaction; the reaction from right to left is the reverse reaction.

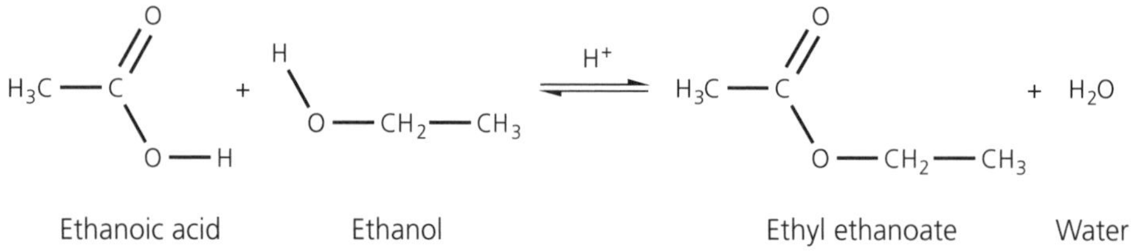

Ethanoic acid   Ethanol   Ethyl ethanoate   Water

Figure 5.15

When ethanoic acid and ethanol are refluxed in the presence of an acid catalyst, the forward reaction, r_1, is initially fast because the concentration of both reagents is high. However, as they react, the concentration of each decreases, lowering the rate of the forward reaction.

The reverse reaction, r_2, is initially very slow because the amounts of ethyl ethanoate and water present are very small. However, as time progresses, the concentration of ethyl ethanoate and water gradually increases, as does the rate of the reverse reaction.

In summary, the forward reaction, r_1, starts off fast but slows down, while the reverse reaction, r_2, starts slowly and speeds up. It follows that they will reach a point when the rate of the forward reaction exactly equals the rate of the reverse reaction:

$$r_1 = r_2$$

When this happens the system is said to be in **dynamic equilibrium**. It is dynamic because the reagents and the products are constantly interchanging. It is in equilibrium because the amount of each chemical

> A **dynamic equilibrium** is reached when the rate of the forward reaction equals the rate of the reverse reaction. The concentration of the reagents and products remain constant; the reagent and the product molecules react continuously.

in the system remains constant. Dynamic equilibrium can only be reached if the system is closed.

Now test yourself

TESTED

9 Explain what is meant by the terms *reversible reaction* and *dynamic equilibrium*.

Answer on p. 117

Le Chatelier's principle

REVISED

Le Chatelier's principle is used to predict the effect of changes in conditions on the position of equilibrium.

The factors that can be readily changed are concentration, temperature and pressure:

- If the concentration is increased, the system will move to decrease the concentration.
- If the temperature is increased, the system will move to decrease the temperature.
- If the pressure is increased, the system will move to decrease the pressure.

Le Chatelier's principle states that if a closed system at equilibrium is subject to a change, the system will move to *minimise* the effect of that change.

Effect of changing pressure on the position of equilibrium

The pressure of a gas mixture depends on the number of gas molecules in the mixture. The greater the number of gas molecules in the equilibrium mixture, the greater is the pressure in the mixture. If the pressure is increased, a system at equilibrium alters to decrease the pressure by reducing the number of gas molecules in the system.

In a system such as $2SO_2(g) + O_2(g) \rightleftharpoons 2SO_3(g)$, if the pressure is increased, the position of the equilibrium moves to the *right*, so that the number of gas molecules is reduced from 3 to 2. This has the effect of reducing the pressure.

If the pressure is increased on a system such as $N_2O_4(g) \rightleftharpoons 2NO_2(g)$, the position of the equilibrium moves to the *left*, so that the number of gas molecules is reduced from 2 to 1. This has the effect of reducing the pressure.

If the pressure is increased on a system such as $2HI(g) \rightleftharpoons H_2(g) + I_2(g)$, the position of the equilibrium does not move, because there are the same number of gas molecules on each side of the equilibrium. A change in the equilibrium position has no effect on the pressure.

Effect of changing temperature on the position of equilibrium

Temperature not only influences the rate of the reaction it also plays an important role in determining the equilibrium position. The effect of temperature can only be predicted if the ΔH value of the reaction is known.

Consider the reaction:

$2A(g) + B(g) \rightleftharpoons C(g) + D(g)$ $\Delta H = -100\,\text{kJ}\,\text{mol}^{-1}$

It follows that:
- the forward reaction: $2A(g) + B(g) \rightarrow C(g) + D(g)$ is exothermic
  ($\Delta H = -100\,\text{kJ mol}^{-1}$)
- the reverse reaction: $C(g) + D(g) \rightarrow 2A(g) + B(g)$ is endothermic
  ($\Delta H = +100\,\text{kJ mol}^{-1}$)

According to le Chatelier's principle, if we increase the temperature of the reaction mixture, the system alters to decrease the temperature. This is achieved by the system favouring the reverse reaction, which is endothermic. This, therefore, helps to remove the additional enthalpy caused by increasing the temperature. The position of the equilibrium moves to the left.

Decomposition of hydrogen iodide is an endothermic reaction:

$$2HI(g) \rightleftharpoons H_2(g) + I_2(g)$$

- If the temperature is increased, the equilibrium moves to the right.
- If the temperature is decreased, the equilibrium moves to the left.

Oxidation of sulfur dioxide is an exothermic reaction:

$$SO_2(g) + O_2(g) \rightleftharpoons 2SO_3(g)$$

- If the temperature is increased, the equilibrium moves to the left.
- If the temperature is decreased, the equilibrium moves to the right.

Now test yourself

TESTED

10 State le Chatelier's principle.
11 Use le Chatelier's principle to deduce what happens to the following equilibrium when it is subjected to the stated changes:

$$2NO_2(g) \rightleftharpoons N_2O_4(g) \quad \Delta H = -57.2\,\text{kJ mol}^{-1}$$

   (a) the temperature is increased
   (b) the pressure is decreased
   (c) the $N_2O_4(g)$ is removed

Answers on p. 113

Effect of a catalyst on the position of equilibrium

A catalyst is a substance that speeds up the rate of reaction by providing an alternative route, or mechanism, that has a lower activation energy. A catalyst does *not* alter the amount of product.

In a system at equilibrium, a catalyst speeds up the forward and the reverse reactions equally. Therefore, a catalyst has no effect on the *position* of the equilibrium but it reduces the time taken for equilibrium to be reached.

Compromise conditions — the Haber process

REVISED

Large quantities of nitrogenous compounds, particularly fertilisers, are needed by humans. Atmospheric nitrogen is in plentiful supply but cannot be used directly. The Haber process is used to 'fix' atmospheric nitrogen and convert it into ammonia:

$$N_2(g) + 3H_2(g) \rightleftharpoons 2NH_3(g) \qquad \Delta H = -93\,\text{kJ mol}^{-1}$$

Le Chatelier's principle allows the optimum conditions for this industrial process to be determined.

The enthalpy change is $-93\,kJ\,mol^{-1}$, so the forward reaction is exothermic. Therefore, if the temperature is decreased the equilibrium moves to the right, so a low temperature is best (optimum) for the formation of ammonia.

If the pressure is increased on the system $N_2(g) + 3H_2(g) \rightleftharpoons 2NH_3(g)$, the equilibrium position moves to the right to reduce the number of molecules. This has the effect of reducing the pressure. Therefore, high pressure is optimum for the formation of ammonia.

The optimum conditions for the maximum yield of ammonia are low temperature and high pressure. However, the operating conditions of a modern plant are:
● a temperature of around 700 K (427°C)
● a pressure of around 100 atm

The conditions are a compromise. The temperature is a compromise between yield and rate — low temperature gives a high yield but the reaction is too slow.

The pressure is a compromise between yield and cost/safety — high pressure gives a high yield but it is both costly and dangerous (hydrogen gas is highly explosive).

The rate of reaction is improved by using a catalyst of finely divided iron or iron in a porous form that incorporates metal oxide promoters.

Table 5.4 gives the boiling points of the gases involved in the Haber process.

Table 5.4 Boiling points of the gases involved in the Haber process

Gas	$N_2(g)$	$H_2(g)$	$NH_3(g)$
Boiling point	77 K (−196°C)	20 K (−253°C)	240 K (−33°C)

If the equilibrium mixture is cooled to about −40°C the ammonia gas liquefies, so ammonia gas is lost from the equilibrium mixture. The system moves to minimise the effect of this loss. The reagents react to produce more ammonia gas to replace the ammonia gas that was liquefied. Any unreacted nitrogen and hydrogen is recycled and used again in a continuous process.

The equilibrium constant, K_c

REVISED

For any reversible reaction that has reached dynamic equilibrium the concentrations of the reactants and products remain unchanged and it is possible to use these concentrations to calculate a numerical value for an important constant, K_c, known as the equilibrium constant. The equilibrium constant varies with temperature but its value is the same for any other change in condition.

In general terms for the reaction:

$aA + bB \rightleftharpoons cC + dD$

the equilibrium constant,

$$K_c = \frac{[C]^c[D]^d}{[A]^a[B]^b}$$

In the expression for K_c the square brackets '[ ]' indicate that the concentrations of the reactants and products are expressed in units of $mol\,dm^{-3}$.

The concentrations of the chemicals on the right-hand side of the equation appear on the top line of the expression. The concentrations of reactants on the left appear on the bottom line. Each concentration term is raised to the power of the number in front of its formula in the balanced equation.

For the equilibrium:

$$3H_2(g) + N_2(g) \rightleftharpoons 2NH_3(g)$$

$$K_c = \frac{[NH_3(g)]^2}{[H_2(g)]^3[N_2(g)]}$$

If there is a solid reactant or product in the equilibrium then it is not included in the expression for K_c.

For the equilibrium:

$$H_2O(g) + C(s) \rightleftharpoons H_2(g) + CO(g)$$

$$K_c = \frac{[H_2(g)][CO(g)]}{[H_2O(g)]}$$

The value of K_c is important because it gives an indication of the balance of reactants and products present in the equilibrium mixture. If K_c has a large value it will mean that there are more of the products present than the reactants, if K_c is small then it is the reactants that must be present in larger quantity. The values of K_c do, in fact, vary enormously from large values down to values that are very small.

For example in the equilibrium:

$$CH_3COOH(l) + C_2H_5OH(l) \rightleftharpoons CH_3COOC_2H_5(l) + H_2O(l)$$

$$K_c = \frac{[CH_3COOC_2H_5(l)][H_2O(l)]}{[CH_3COOH(l)][C_2H_5OH(l)]}$$

The value of K_c is approximately 4 at room temperature so there is approximately four times as much product as there are reactants. K_c will not have any units because the concentrations of each component will cancel out in the equilibrium expression.

For the equilibrium between sulfur dioxide, oxygen and sulfur trioxide:

$$2SO_2(g) + O_2(g) \rightleftharpoons 2SO_3(g)$$

the equilibrium constant

$$K_c = \frac{[SO_3(g)]^2}{[SO_2(g)]^2[O_2(g)]}$$

has a size of approximately 1.0×10^{12} at 230°C so in this case the equilibrium mixture has vastly more of the product, SO_3, than the reactants. In this case K_c will have units of $1/mol\,dm^{-3}$ or $dm^{+3}\,mol^{-1}$ because the concentrations do not completely cancel in the equilibrium expression.

> **Exam tip**
>
> You will not be expected to supply units for equilibrium constants in the AS exam.

The effect on K_c of a change in temperature

The effect of a change in temperature on the value of K_c is best explained using an example.

The equilibrium:

$$2SO_2(g) + O_2(g) \rightleftharpoons 2SO_3(g) \qquad \Delta H = -197\,kJ\,mol^{-1}$$

is exothermic. An increase in temperature moves the equilibrium to the left and less SO_3 is produced. It must therefore follow that the value of the equilibrium constant will decrease as the temperature rises.

Calculating an equilibrium constant

If the equilibrium concentration of each component of an equilibrium mixture is known, the equilibrium constant can be calculated by substituting these values into the expression for the equilibrium constant.

Example

At 450°C, hydrogen and gaseous iodine form an equilibrium mixture with hydrogen iodide:

$$H_2(g) + I_2(g) \rightleftharpoons 2HI(g)$$

When an equilibrium mixture is analysed it is found to contain $0.015\,mol\,dm^{-3}$ of HI, $0.0012\,mol\,dm^{-3}$ of I_2 and $0.0038\,mol\,dm^{-3}$ of H_2.

Calculate the value of the equilibrium constant to 2 significant figures.

Answer

$$K_c = \frac{[HI(g)]^2}{[H_2(g)][I_2(g)]}$$

Therefore:

$$K_c = \frac{(0.015)^2}{(0.0012)(0.0038)} = 49$$

Now test yourself

12 Write the equilibrium constant for the equilibrium:

$$2NO_2(g) \rightleftharpoons N_2O_4(g)$$

   (a) At a certain temperature the equilibrium constant for this reaction has a value of 0.0025. What does this indicate about the position of the equilibrium?
   (b) If the concentration of NO_2 is $1.0\,mol\,dm^{-3}$, calculate the concentration of the N_2O_4 at this temperature.
13 When the equilibrium $2NOCl(g) \rightleftharpoons 2NO(g) + Cl_2(g)$ is analysed it is found that the equilibrium concentrations of the gases are $NOCl = 3.42\,mol\,dm^{-3}$, $NO = 0.32\,mol\,dm^{-3}$, $Cl_2 = 0.16\,mol\,dm^{-3}$. Calculate K_c for the equilibrium.

Answers on p. 117

Exam practice

1 Bond enthalpies can provide information about the energy changes that accompany a chemical reaction.
   (a) What do you understand by the term *bond enthalpy*? [2]
   (b) (i) Write an equation, including state symbols, to illustrate the bond enthalpy of hydrogen chloride. [1]
       (ii) Write an equation to illustrate the bond enthalpy of methane. [2]
   (c) (i) The table below shows some average bond enthalpies.

Bond	Average bond enthalpy/$kJ\,mol^{-1}$
C–C	350
C=C	610
H–H	436
C–H	410

Use the information in the table to calculate the enthalpy change for the hydrogenation of ethene (Figure 5.16). [3]

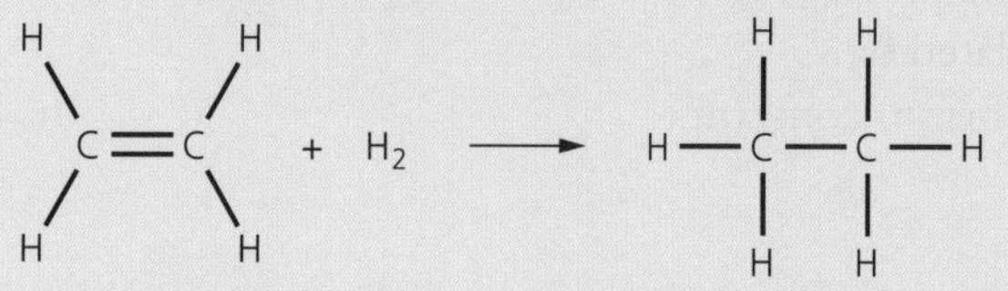

Ethene (g)   Ethane (g)

Figure 5.16

    (ii) The enthalpy change of this reaction is found by experiment to be $-136\,kJ\,mol^{-1}$. Explain why this value is different from that calculated in (i). [2]

  (d) In the hydrogenation of ethane, nickel is used as a catalyst. Explain the mode of action of nickel in this reaction. [3]

2 When a mixture of hydrogen and oxygen is ignited by a spark, water is produced:

$$2H_2(g) + O_2(g) \rightarrow 2H_2O(l) \quad \Delta H = -571.6\,kJ\,mol^{-1}$$

$$2H_2(g) + O_2(g) \rightarrow 2H_2O(g) \quad \Delta H = -483.6\,kJ\,mol^{-1}$$

  (a) On the same diagram, draw the enthalpy profiles for the formation of water and steam. [2]

  (b) Use the enthalpy profile diagram to deduce the enthalpy change: [2]

$$H_2O(l) \rightarrow H_2O(g)$$

3 Octane, C_8H_{18}, is one of the hydrocarbons present in petrol.

  (a) Define the term standard enthalpy change of combustion. [3]

  (b) Use the data below to calculate the standard enthalpy change of combustion of octane. [3]

$$C_8H_{18}(l) + 12\tfrac{1}{2}O_2(g) \rightarrow 8CO_2(g) + 9H_2O(l)$$

Compound	$\Delta_f H^{\ominus}/kJ\,mol^{-1}$
$C_8H_{18}(l)$	−250.0
$CO_2(g)$	−393.5
$H_2O(l)$	−285.9

  (c) Combustion in car engines produces polluting gases, mainly carbon monoxide, unburnt hydrocarbons and oxides of nitrogen such as nitrogen(II) oxide, NO. Explain, with the aid of equations, how carbon monoxide and nitrogen(II) oxide are produced in a car engine. [2]

  (d) (i) The catalytic converter removes much of this pollution in a series of reactions. Write an equation to show the removal of carbon monoxide and nitrogen(II) oxide gases. [1]

    (ii) The removal of carbon monoxide and nitrogen(II) oxide gases involves a redox reaction. Use your answer to d(i) to identify the element being reduced and state the change in its oxidation number. [2]

4 Sulfuric acid, H_2SO_4, is made industrially by the contact process. This is an example of a dynamic equilibrium:

$$2SO_2(g) + O_2(g) \rightleftharpoons 2SO_3(g) \quad \Delta H = -98\,kJ\,mol^{-1}$$

  (a) State *two* features of a reaction with a *dynamic equilibrium*. [2]

  (b) State and explain what happens to the equilibrium position of the above reaction when: [6]

    (i) the temperature is raised

    (ii) the pressure is increased

    (iii) Suggest the optimum conditions for the contact process.

  (c) (i) The conditions used for the contact process are a temperature of between 450°C and 600°C and a pressure of about 10 atm. Explain why the optimum conditions are not used. [3]

    (ii) Vanadium(v) oxide is used as a catalyst. What effect does this have on the conversion of $SO_2(g)$ into $SO_3(g)$? [2]

    (iii) At least three catalyst chambers are used to ensure maximum conversion of $SO_2(g)$. The conversion yield can exceed 98%. State two advantages of this high conversion rate. [2]

5 In the Haber process, hydrogen is reacted with nitrogen in the presence of a catalyst to produce ammonia:

$$N_2(g) + 3H_2(g) \rightleftharpoons 2NH_3(g)$$

The activation energy for the forward reaction is $+68 \, kJ \, mol^{-1}$.
The activation energy for the reverse reaction is $+160 \, kJ \, mol^{-1}$.

(a) (i) Use this information to sketch the energy profile diagram. Label clearly the activation energy for the forward reaction, E_f, and the activation energy for the reverse reaction, E_r. [2]

(ii) Explain what is meant by activation energy. [1]

(iii) Calculate the enthalpy change for the forward reaction. [1]

(b) Much of the ammonia produced is oxidised into nitric acid using the Ostwald process, which involves three stages:

Stage 1	$4NH_3(g) + 5O_2(g) \rightleftharpoons 4NO(g) + 6H_2O(g)$	$\Delta H = -950 \, kJ \, mol^{-1}$
Stage 2	$2NO(g) + O_2(g) \rightleftharpoons 2NO_2(g)$	$\Delta H = -114 \, kJ \, mol^{-1}$
Stage 3	$3NO_2(g) + H_2O(g) \rightleftharpoons 2HNO_3(g) + NO(g)$	$\Delta H = -117 \, kJ \, mol^{-1}$

(i) With reference to the oxidation number of the nitrogen in $NH_3(g)$ (Stage 1) and in HNO_3 (Stage 3) show that this is an oxidation process. [3]

(ii) State le Chatelier's principle. [1]

(iii) Use le Chatelier's principle to predict and explain the temperature and the pressure that would give the maximum yield at equilibrium in Stage 1. [4]

(iv) Suggest what happens to the NO(g) produced in Stage 3. [1]

(c) The nitric acid produced in Stage 3 is a strong acid. Explain, with the aid of an equation, what is meant by the term *strong acid*. [2]

6 In an investigation to find the enthalpy change of combustion of ethanol, C_2H_5OH, a student found that 1.60 g of ethanol could heat 150 g of water from 22.0°C to 71.0°C. The specific heat capacity of ethanol is $4.2 \, J \, g^{-1} \, K^{-1}$.

(a) Use the results to calculate a value for the enthalpy change of combustion of ethanol. [6]

(b) The theoretical value of the standard enthalpy change of combustion of ethanol is $-1367.3 \, kJ \, mol^{-1}$. Give two reasons for the difference between the theoretical and experimental values. Suggest an improvement that could be made to the experiment to minimise the most significant error. [3]

(c) Catalysts are of great economic importance. Give an example of a catalyst that is in the same state as the reactants and an example of a catalyst that is in a different state from the reactants. State why the reactions you have chosen are important and explain how a catalyst increases the rate of reaction. [6]

7 When the indicator methyl orange is dissolved in water the following dynamic equilibrium is set up.

Yellow   **Red**

(a) Hydrochloric acid is added to the equilibrium mixture. State the colour change you would see. Explain your answer.

(b) Aqueous potassium hydroxide solution is added dropwise to the solution in **a** until no further colour change occurs. Suggest all the colour changes you would see. Explain your answer. [6]

Answers and quick quiz 5 online

ONLINE

Summary

You should now have an understanding of:
- exothermic and endothermic reactions
- enthalpies of reaction, formation and combustion
- bond enthalpies
- Hess's law and calculations using enthalpy cycles

- Boltzmann distributions and how temperature and catalysts affect rate
- catalysts
- le Chatelier's principle
- the equilibrium constant, K_c

6 Basic concepts and hydrocarbons

Naming and formulae of organic compounds

Homologous series

REVISED

Organic compounds are grouped together in families called **homologous series**.

You are expected to be able to recognise and name alkanes, alkenes, alcohols and haloalkanes.

> A **homologous series** is a group of compounds that have the same general formula and contain the same functional group (and therefore have similar chemical properties). Each member of the series differs from the next by CH_2.

Table 6.1 Formulae and names of hydrocarbons

	General formula	Naming
Alkanes	C_nH_{2n+2}	Name always ends '-ane'
Alkenes	C_nH_{2n}	Name always ends '-ene
Alcohols	$C_nH_{2n+1}OH$	Name always ends '-ol'
Haloalkanes	$C_nH_{2n+1}X$ where X = Cl, Br, I	Name always starts 'chloro-', 'bromo-' or 'iodo-'

Figure 6.1 shows how to name organic compounds.

A

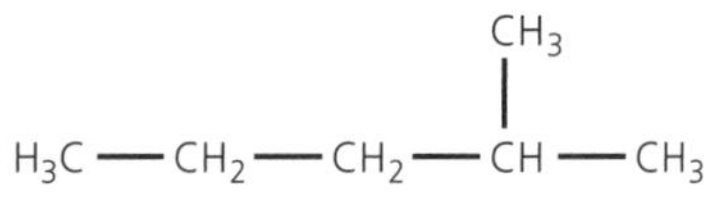

Compound A is an alkane — the name ends in -*ane*
Longest carbon chain is 5 — *pentane*
Branch on the second carbon contains one carbon — hence *2-methyl*
Full name is **2-methylpentane**

B

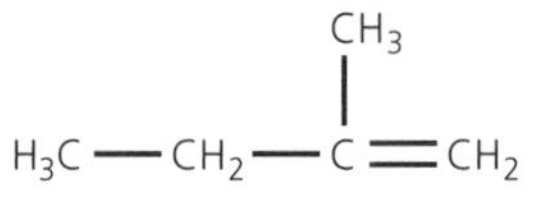

Compound **B** is an alkene — the name ends in -*ene*
Longest carbon chain is 4 — *butene*
Double bond starts at carbon atom 1 — but-*1*-ene
Branch on the second carbon contains one carbon — hence *2-methyl*
Full name is **2-methylbut-1-ene**

C

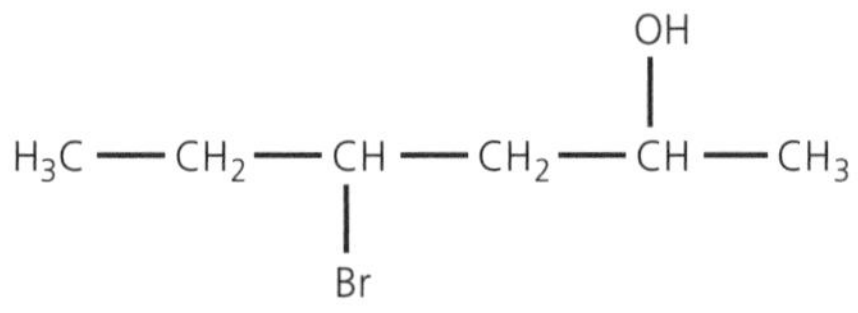

Compound **C** is both an alcohol (ends in -*ol*) and a bromoalkane (starts *bromo*-)
Longest carbon chain is 6 — *hexane*
Alcohol is on the second carbon — hence -*2-ol*
Bromine is on fourth carbon — hence *4-bromo*-
Full name is **4-bromohexan-2-ol**

Figure 6.1 Naming organic compounds

Representing organic formulae

REVISED

You are expected to draw and represent compounds in a number of different ways:

- A **structural formula** is accepted as the minimal detail, using conventional groups, for an unambiguous structure. The structural formula for butane, C_4H_{10}, is $CH_3CH_2CH_2CH_3$.

- A **displayed formula** shows both the relative placing of atoms and the number of bonds between them. The displayed formula of methylpropane is shown in Figure 6.2.

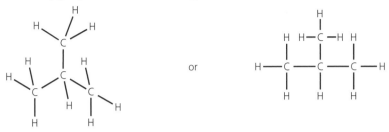

or

Figure 6.2

- A **skeletal formula** is a simplified organic formula. The hydrogen atoms in alkyl chains are not shown, leaving only the carbon skeleton and associated functional groups. The skeletal formulae for butane, methylpropane, pentan-2-ol and but-1-ene are shown in Figure 6.3.

Butane   Methylpropane   Pentan-2-ol   But-1-ene

Figure 6.3

Figure 6.4 shows how cyclic compounds such as cyclohexane and benzene are represented.

Aliphatic

Cyclobutane   Cyclopentane   Cyclohexane

Aromatic

or

Benzene

Figure 6.4

Compounds that contain a benzene ring are **aromatic** compounds whilst compounds that contain a non-aromatic ring are **alicyclic** compounds (Figure 6.5).

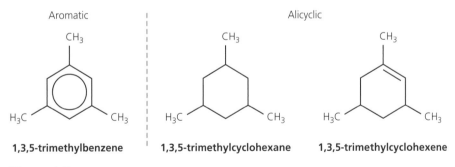

Aromatic

Alicyclic

1,3,5-trimethylbenzene   **1,3,5-trimethylcyclohexane**   **1,3,5-trimethylcyclohexene**

Figure 6.5

Now test yourself

TESTED

1 Draw the displayed formula for each of the following molecules: 2-chloropropane, 1-chloropropane, butan-2-ol, 2-methylpentane, 3-methylbut-1-ene

2 Name each of the following:

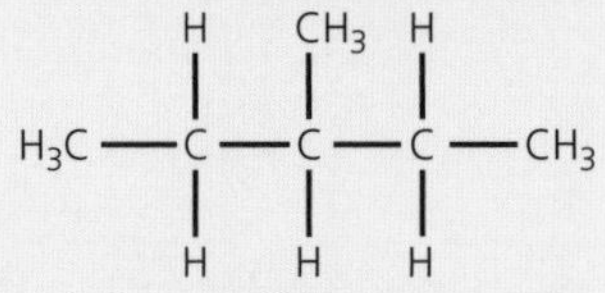

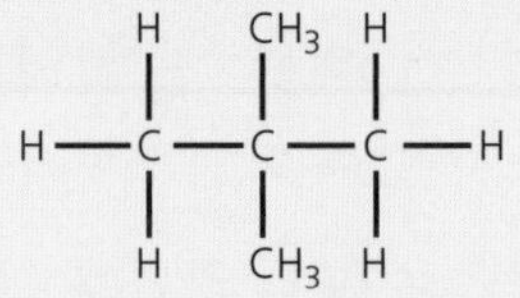

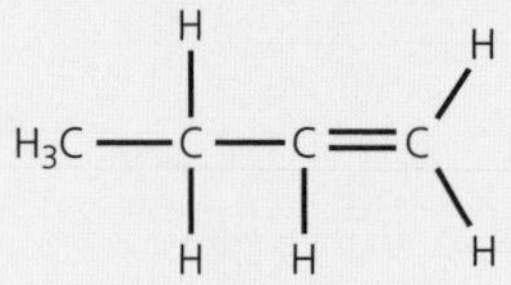

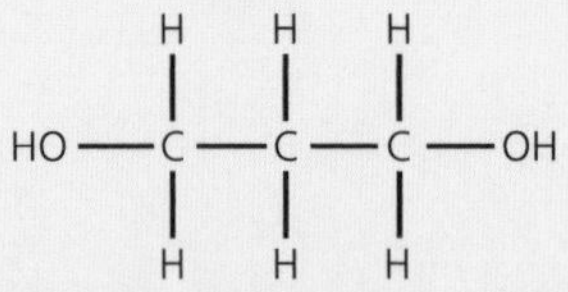

3 Name the following: $CH_3CHBrCH_3$, $CH_3CHCHCH_3$, $(CH_3)_4C$ and $CH_3CH_2CH(CH_3)CH_2CH_3$.

Answers on p. 117

Isomerism

Structural isomerism

REVISED

Structural isomers with molecular formula $C_4H_{10}O$ are shown in Figure 6.6.

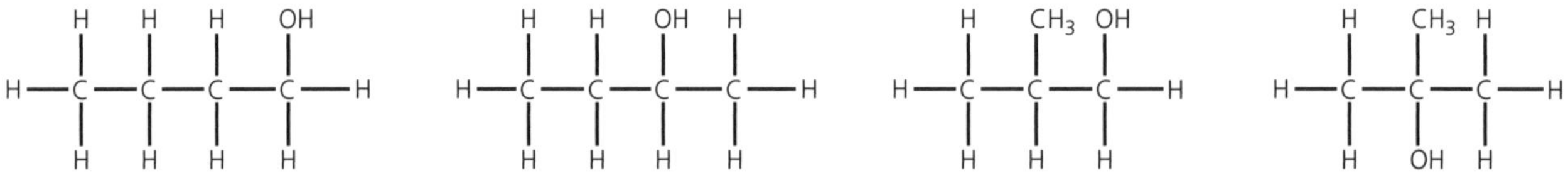

Figure 6.6 Structural isomers of $C_4H_{10}O$

> **Structural isomers** are compounds that have the same molecular formula but different structural formulae.

Questions are sometimes of the following form:

There are five isomers of C_6H_{14}. Three are drawn for you (Figure 6.7). Draw the other two.

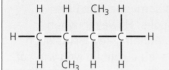 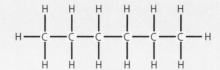 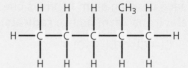

Figure 6.7

Few students gain both marks. Most redraw at least one of the above three isomers, but in a different orientation.

Skeletal isomers also cause problems — for example, when asked to draw isomers such as 1-chloropropane.

Many draw the carbon skeleton first as:

which correctly represents propane.

Most then add Cl to give:

which changes it to chloroethane.

The correct skeletal formula for 1-chloropropane is:

E/Z isomerism (geometric isomerism) REVISED

E/Z **isomerism** is found in alkenes. The key features to look for are:
- the C=C double bond
- each carbon atom in the C=C double bond being bonded to two different atoms or groups

E/Z isomerism is explained fully in the section on alkenes.

> E/Z **isomers** have the same structural formula but the atoms are arranged differently in space around a C=C double bond.

Reactions of functional groups REVISED

When describing the reactions of any functional group you are expected to know the reagents, the conditions and to be able to write both a balanced equation and the mechanism.

Reagents These are the chemicals involved in the reaction.

Conditions These normally describe the temperature, pressure and/or the use of a catalyst,

Mechanism The overall reaction is broken down into separate steps. It is usual to identify the attacking species. This can be a **radical**, an **electrophile** or a **nucleophile**.

> A **radical** is a reactive particle with an unpaired electron. The symbol for a radical shows the unpaired electron as a dot, e.g. Cl•, CH_3•.

> An **electrophile** is a reactive ion or molecule that attacks an electron-rich part of a molecule to form a new covalent bond. An electrophile is an **electron-pair acceptor**. Examples include H^+, NO_2^+.

> A **nucleophile** is a molecule or ion with a lone pair of electrons that can form a new covalent bond. A nucleophile is an **electron-pair donor**. Examples include OH^-, $:NH_3$.

The covalent bonds within organic compounds can be broken by either **homolytic fission** or by **heterolytic fission**.

Exam tip

Words that end with '-phile' indicate a liking for — for example, a 'bibliophile' is someone who likes books. However, when explaining key terms such as 'electrophile' you will *not* score any marks for writing that 'it loves electrons'. Stick to the scientific definitions given above.

Homolytic fission occurs when a covalent bond is broken and each bonding atom receives one electron from the bonding pair of electrons, forming two **radicals**:

$$Cl—Cl \rightarrow Cl· + ·Cl$$

Heterolytic fission occurs when a covalent bond is broken and one bonding atom receives both electrons from the bonding pair of electrons, forming a **nucleophile** and an **electrophile**:

$$Cl—Cl \longrightarrow Cl^+ + :Cl^-$$

Electrophile   Nucleophile

The curly arrow shows the movement of an electron pair, and shows heterolytic fission and formation of a covalent bond.

Alkanes

Physical properties of alkanes

REVISED

Hydrocarbons are compounds that contain hydrogen and carbon only. Alkanes and cycloalkanes are saturated hydrocarbons, as all of the C—C bonds are single bonds. This results in a tetrahedral shape, with bond angles 109.5°, around each carbon atom.

Different alkanes have different boiling points. The variation in boiling points depends on the amount of intermolecular forces. Alkanes have very low bond polarity and so the only type of intermolecular force is induced dipole–dipole interactions (van der Waals forces). There are two important trends in the variation of boiling points. First, as the relative molecular mass increases, the boiling point increases. This is due to

● an increase in chain length
● an increase in the number of electrons

Both of the above result in an increase in the number of induced dipole–dipole interactions. Second, for isomers with the same relative molecular mass, the boiling points decrease with an increase in the amount of branching. This can be explained by the fact that straight chains pack closer together, creating more intermolecular forces. This can be illustrated by the three isomers of pentane, C_5H_{12} (Figure 6.8).

Pentane: boiling point = 36°C

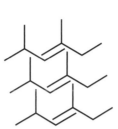

Pack together easily

Methylbutane: boiling point = 28°C

Branched isomers cannot pack together easily and hence have less surface interaction

Dimethylpropane: boiling point = 10°C

Figure 6.8

Chemical properties of alkanes

Alkanes are relatively unreactive because:
- they have very low bond polarity
- the C−C and the C−H bonds are strong bonds

Combustion of alkanes

Alkanes burn easily, releasing energy (exothermic). They are used as fuels in industry, the home and in transport.

Complete combustion of alkanes in an excess of oxygen produces carbon dioxide and water.

$$C_2H_6 + 3\tfrac{1}{2}O_2 \rightarrow 2CO_2 + 3H_2O$$

Incomplete combustion of alkanes in a limited supply of oxygen produces carbon monoxide and water.

$$C_2H_6 + 2\tfrac{1}{2}O_2 \rightarrow 2CO + 3H_2O$$

Carbon monoxide is poisonous, so it is essential that hydrocarbon fuels are burnt in a plentiful supply of oxygen. Cars are fitted with catalytic converters to ensure that the amount of carbon monoxide emitted is reduced.

Reactions of alkanes

When describing the reactions of any functional group you are expected to know the reagents, the conditions and to be able to write both a balanced equation and the mechanism.

Reaction between methane and bromine

Reagent: bromine

Conditions: ultraviolet light

Equation: $CH_4 + Br_2 \rightarrow CH_3Br + HBr$

Mechanism: radical substitution

Initiation: $Br_2 \rightarrow 2Br\bullet$

Propagation 1: $CH_4 + Br\bullet \rightarrow HBr + \bullet CH_3$

Propagation 2: $\bullet CH_3 + Br_2 \rightarrow CH_3Br + Br\bullet$

Termination: $\bullet CH_3 + \bullet CH_3 \rightarrow C_2H_6$ or $\bullet CH_3 + Br\bullet \rightarrow CH_3Br$

There are three distinct stages to the mechanism:

1 Initiation — radicals are generated. The ultraviolet light provides sufficient energy to break the Br−Br bond **homolytically** and generates radicals:

$Br_2 \rightarrow 2Br\bullet$

2 Propagation — involves two steps, each of which maintains the radical concentration. Usually a chlorine or bromine radical is swapped for an alkyl radical or vice versa (Figure 6.9).

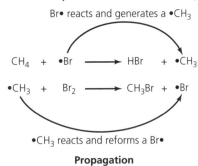

Br• reacts and generates a •CH₃

$CH_4 + \bullet Br \longrightarrow HBr + \bullet CH_3$

$\bullet CH_3 + Br_2 \longrightarrow CH_3Br + \bullet Br$

•CH₃ reacts and reforms a Br•

Propagation

Figure 6.9

3 Termination — involves the loss of radicals:

$\bullet CH_3 + \bullet CH_3 \rightarrow C_2H_6$

Radical reactions have limitations. It is almost impossible to produce a single product because radicals are very reactive and it is difficult to avoid multiple substitutions of the hydrogen atoms in the alkane. For the reaction between CH_4 and Cl_2 the product would typically contain CH_3Cl, CH_2Cl_2, $CHCl_3$ and CCl_4. This makes separation difficult and costly.

Now test yourself

4 Calculate the formulae of alkanes **A** and **B**.
  (a) 100 cm³ of alkane **A** was burnt in excess $O_2(g)$. It produced 300 cm³ $CO_2(g)$ and 400 cm³ $H_2O(g)$.
  (b) When burnt in excess $O_2(g)$, 0.1 mol of alkane **B** produced 10.8 g $H_2O(l)$.
5 Write the balanced equation and the mechanism for the reaction between cyclobutane and chlorine. Explain each of the following terms: radical; homolytic fission; initiation; propagation; termination and substitution.

Answers on p. 117

Alkenes

Alkenes and cycloalkenes are unsaturated hydrocarbons and contain at least one C=C double bond. The double bond is made up of σ-bonds and a π-bond.

> A σ-**bond** is a single covalent bond made up of two shared electrons with the electron density concentrated between the two nuclei.
>
> A π-**bond** is formed by the sideways overlap of two adjacent p-orbitals (Figure 6.10).

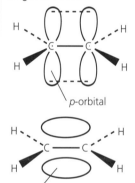

π molecular orbitals above and below the plane of the molecule

Figure 6.10

Typical mistake

When asked to describe, with the aid of a diagram, how the π-bond is formed in an alkene, students often draw the following, which shows a treble (not double) bond between the two carbons:

The correct response is:

The bond angle on each side of the C=C double bond is approximately 120° (usually in the region 116–124°) which results in a trigonal planar structure (Figure 6.11).

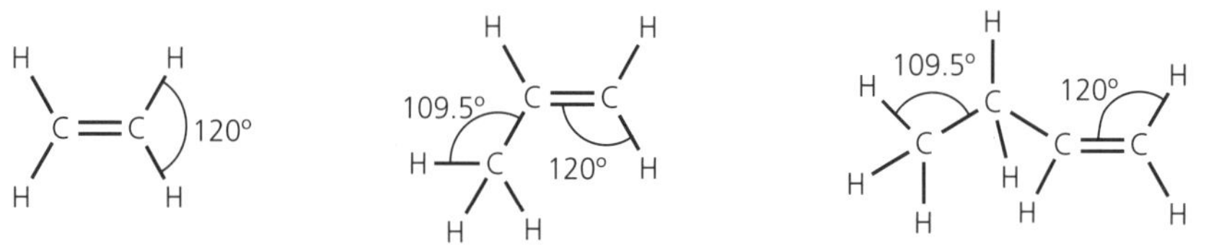

Figure 6.11

The C=C double bond prevents freedom of rotation, which under certain circumstances can lead to the existence of E/Z isomers.

The C=C double bond ensures that there is restricted rotation about the bond and the different atoms or groups attached to each carbon atom ensure that there is no symmetry around the carbon atoms in the C=C double bond (Figure 6.12).

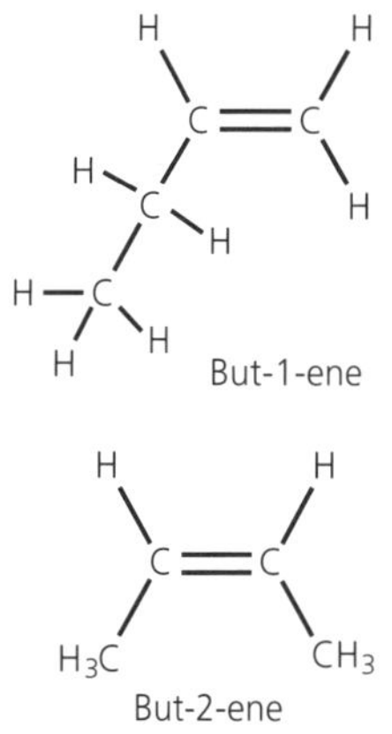

But-1-ene and but-2-ene both have a C=C double bond but the right-hand carbon atom in the C=C double bond in but-1-ene is bonded to two hydrogen atoms and therefore does not exhibit E/Z isomerism. But-2-ene possesses both essential key features and hence has a Z- and an E-isomer (Figure 6.13).

But-1-ene

But-2-ene

Figure 6.12

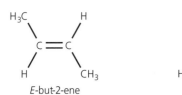

E/Z isomers have a different geometry or 3D shape: this can be seen clearly when they are drawn as skeletal formulae

E-but-2-ene

Z-but-2-ene

Figure 6.13

Cahn, Ingold and Prelog rules

The Cahn, Ingold and Prelog (CIP) rules enable you to decide whether you should name each isomer as an *E*-isomer or *a Z*-isomer. The atomic number of each atom bonded to the C in the C=C double bond determines whether or not it is an *E* isomer or a *Z* isomer:
- If the two attached atoms with the highest atomic numbers are on the diagonally opposite sides of the double bonds it is an *E*-isomer.
- If the two attached atoms with the highest atomic numbers are *not* diagonally opposite each other across the double bonds it is a *Z*-isomer.

Example

In a compound like $CH_3(Cl)C=CHOH$ (Figure 6.14), the C on the left-hand side of the C=C double bond is bonded to a Cl (atomic number 17) and a C (atomic number 6), while the C on the right-hand side is bonded to an H (atomic number 1) and an O (atomic number 8).

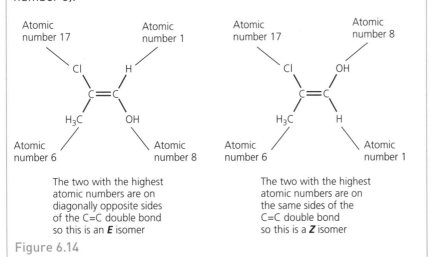

Figure 6.14

- If two attached atoms have the same atomic number then the adjacent atoms with the highest atomic number are taken into account. This occurs with attached alkyl groups such that $CH_3CH_2CH_2 > CH_3CH_2 > CH_3$.

In Figure 6.15 the C atom on the right of the C=C bond is bonded to two carbon atoms, each with atomic number 6. It is still possible to have *E/Z* isomers by considering the adjacent atoms: CH_3 has a mass of 15 and CH_3CH_2 has a mass of 29. It follows that the isomer on the left is the *Z*-stereoisomer and the isomer on the right is the *E*-isomer.

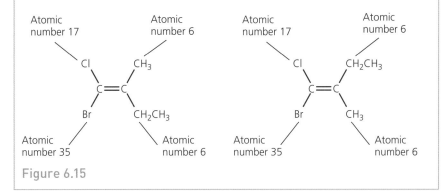

Figure 6.15

Cis-trans isomerism is a special case of *E/Z* isomerism in which two of the substituent groups attached to each carbon atom of the C=C group are the same.

Now test yourself

TESTED ☐

6 Draw and name the isomers of C_6H_{14}. Explain which isomers have the lowest boiling point.
7 Explain, using compounds of C_4H_8, what is meant by isomerism. Include in your answer structural isomerism and *E/Z* isomerism.

Answers on pp. 117 and 118

Addition reactions of alkenes

REVISED ☐

The C=C double bond is an unsaturated bond and therefore undergoes addition reactions. Essentially, the double bond opens and an atom or group adds to each of the carbons. The general reaction can be summarised as shown in Figure 6.16.

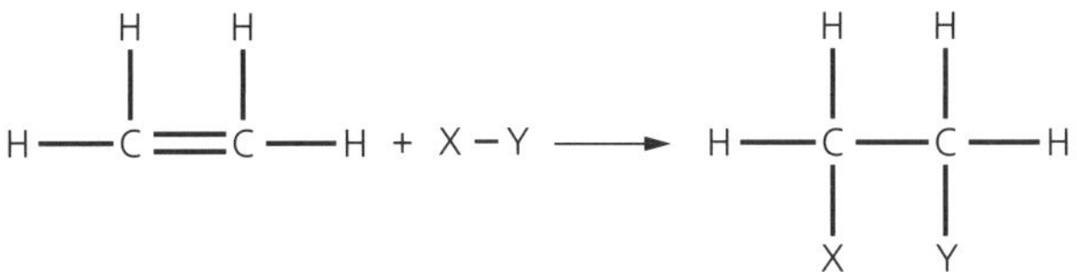

$$H-\overset{\overset{\displaystyle H}{|}}{C}=\overset{\overset{\displaystyle H}{|}}{C}-H + X-Y \longrightarrow H-\overset{\overset{\displaystyle H}{|}}{\underset{\underset{\displaystyle X}{|}}{C}}-\overset{\overset{\displaystyle H}{|}}{\underset{\underset{\displaystyle Y}{|}}{C}}-H$$

Figure 6.16

When preparing for examinations it is good practice to try to stick to a routine. For most organic reactions it is useful to know:

● reagents
● conditions (if any)
● observations (if any)
● balanced equations

The reactions of ethene are summarised in Figure 6.17.

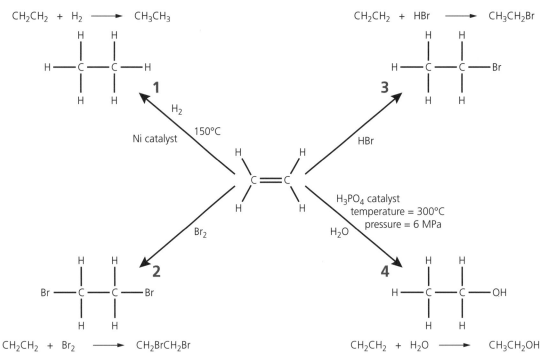

Figure 6.17

In the flow diagram:

- Reaction 1 is **hydrogenation**. It is used to produce margarine by reacting polyunsaturated vegetable oils, derived from plants, with hydrogen.
- Reaction 2 is **bromination**. It is used as a test for alkenes. The brown colour of the bromine fades and the reaction mixture becomes colourless.
- Reaction 3 is the **formation of haloalkanes**. The HBr is made *in situ* from NaBr + H_2SO_4.
- Reaction 4 is **hydration**. Ethene reacts with steam in the presence of a suitable acid catalyst to produce ethanol.

All other alkenes undergo similar reactions under similar conditions.

> **Typical mistake**
>
> When asked to describe what you would see when bromine reacts with an alkene, many students lose the mark by stating that the bromine would go 'clear'. Clear is the wrong word — bromine is already clear, it is a clear red-brown liquid. When it reacts it loses its colour (becomes decolorised).

> **Typical mistake**
>
> When drawing alcohols you have to be careful how you draw the bond to the hydroxyl (OH) group. It is easy to lose the mark, as shown in Figure 6.18a and 6.18b. Figure 6.18c shows how it should be drawn.
>
> **Figure 6.18**

Now test yourself

8 (a) Write a balanced equation for each of the following:
  (i) the *complete* combustion of propene
  (ii) the reaction between but-1-ene and hydrogen
  (iii) hydration of but-2-ene
  (b) Write an equation for the reaction between buta-1,3-diene and bromine. State what you would observe and name the organic product.
  (c) Explain why the hydration of but-2-ene gives only one organic product but the hydration of but-1-ene gives two organic products.

Answer on p. 118

Electrophilic addition

Mechanisms involving electrophiles and nucleophiles involve the movement of electron pairs. This movement is shown by the use of curly arrows. The curly arrow always points from an area that is electron rich to an area that is electron deficient.

When describing mechanisms it is essential that you show:
- relevant dipoles
- lone pairs
- curly arrows

Alkenes, such as ethene, undergo electrophilic addition reactions. An electrophile is an electron pair acceptor that results in the formation of a covalent bond.

The mechanism of the reaction between ethene and bromine is shown in Figure 6.19.

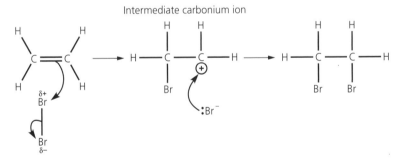

Figure 6.19

The key features of the mechanism are:
1 When the Br—Br approaches the ethene a temporary induced dipole is formed resulting in $Br^{\delta+}$—$Br^{\delta-}$.
2 The initial curly arrow starts at the π-bond (within the C=C double bond) and points to the $Br^{\delta+}$.
3 The second curly arrow shows the movement of the bonded pair of electrons in the Br—Br to the $Br^{\delta-}$ resulting in **heterolytic fission** of the Br—Br bond.
4 The formation of an intermediate carbonium ion (also called a carbocation) and a :Br⁻ ion that now has the pair of electrons that were in the Br—Br bond occurs.
5 The third curly arrow from the :Br⁻ to the positively charged carbonium ion results in the formation of 1,2-dibromoethane.

> **Heterolysis** or **heterolytic fission** occurs when a covalent bond is broken (fission). One of the atoms receives both of the electrons from the covalent bond. This results in the formation of two oppositely charged ions. An example is Cl–$Cl \rightarrow Cl^+ + Cl^-$

Now test yourself

9 (a) Explain what is meant by an *electrophile*.

(b) Write a balanced equation for the reaction between Br_2 and cyclohexene and give the mechanism. Use curly arrows to show the movement of electrons. Show any relevant dipoles and lone pairs of electrons.

(c) When bromine reacts with an alkene, the Br–Br bond undergoes heterolytic fission (Figure 6.20).

$$Br \longrightarrow Br \longrightarrow Br^+ + :Br^-$$

Heterolytic fission

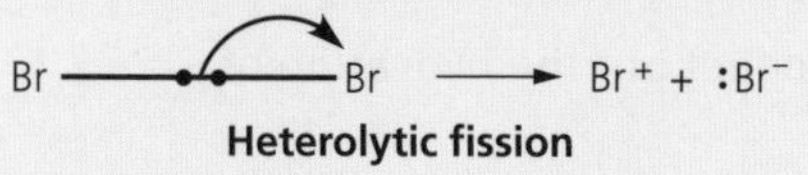

Figure 6.20

Compare this with the way the Br–Br bond is broken when bromine reacts with an alkane.

Answer on p. 118

Exam tip

Mechanisms are generally well understood, but it is easy to lose marks by rushing and/or being careless. Look at the mechanism for the reaction between propene and bromine (Figure 6.21). At first glance it looks good, but there are *seven* errors or omissions. Can you spot all seven?

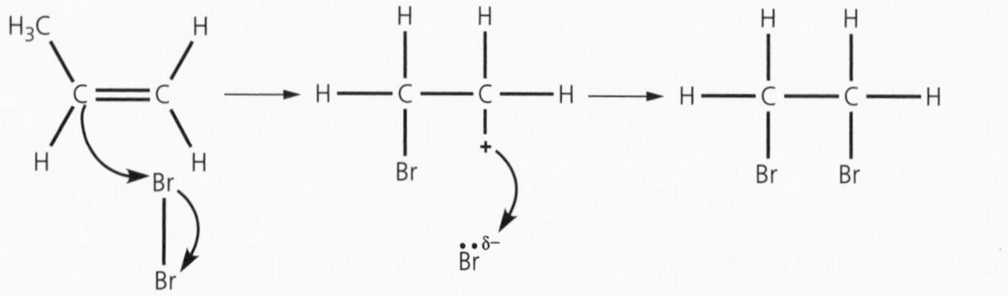

Figure 6.21

If you spot the mistakes made by others, it should prevent you from making the same mistakes.

When reacted with either HBr or H_2O, **unsymmetrical** alkenes, for example propene, produce a mixture of two isomers:

- Reaction of propene with HBr:

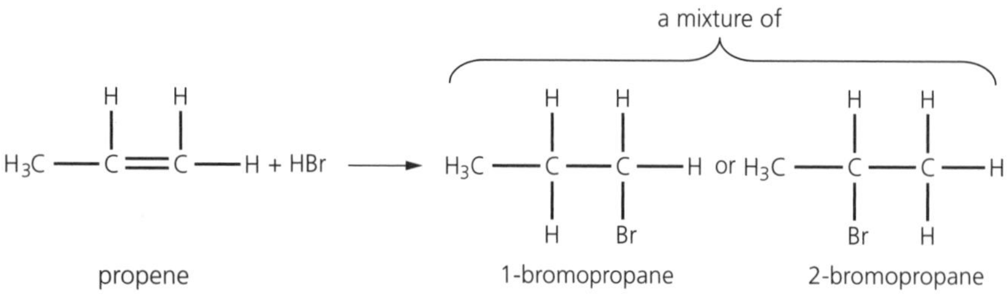

propene   1-bromopropane   2-bromopropane

- Reaction of propene with H_2O:

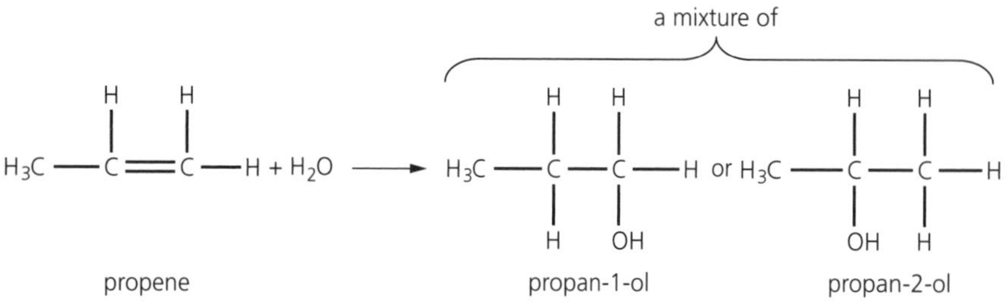

propene   propan-1-ol   propan-2-ol

Exam practice answers and quick quizzes at **www.hoddereducation.co.uk/myrevisionnotes**

When either HBr or H_2O reacts with an unsymmetrical alkene, the major product can be predicted using **Markownikoff's rule**.

Markownikoff's rule states that the addent other than hydrogen goes to the least hydrogenated carbon.

For example, if propene, $CH_3CH=CH_2$, reacts with HBr, the *addent other than hydrogen* is Br and this will bond to the C in the C=C double bond with the lowest number of Hs. The Br will therefore bond to the CH (it only has one H) and not the CH_2 (Figure 6.22).

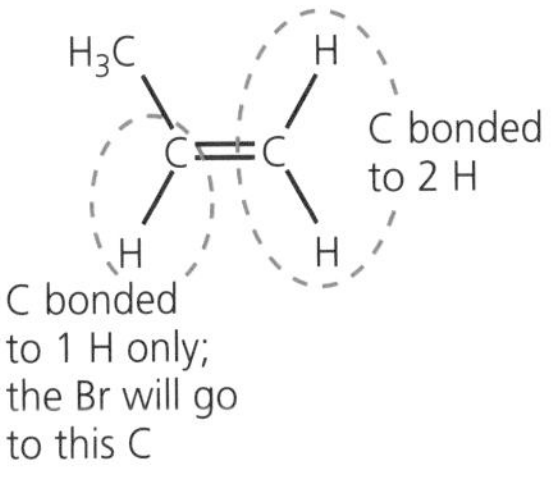

Figure 6.22

Considering the mechanism in detail, there are two alternatives (Figure 6.23).

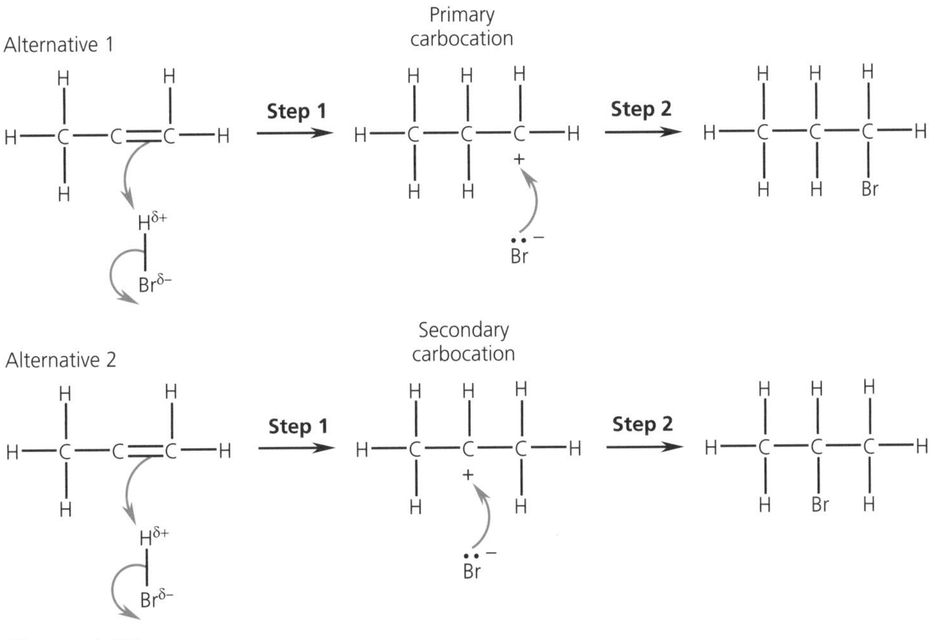

Figure 6.23

The product depends on the relative stabilities of the carbocation intermediates in the mechanism:

Most stable	tertiary carbocation	>	secondary carbocation	>	primary carbocation	Least stable

The more stable the carbocation intermediate, the more likely it will result in the product such that 'alternative 2' in Figure 6.23 is the favoured mechanism, forming 2-bromopropane.

Addition polymerisation of alkenes

REVISED

Alkenes can undergo an addition reaction in which one alkene molecule joins to others and a long molecular chain is built up. The individual alkene molecule is a **monomer** and the long-chain molecule is called a **polymer**.

Some common monomers and their reactions are shown in Figure 6.24. *n* represents a large number and can be as big as 10 000.

Revision activity

On two postcards use but-2-ene and but-1-ene to devise two spider diagrams for the reactions of alkenes.

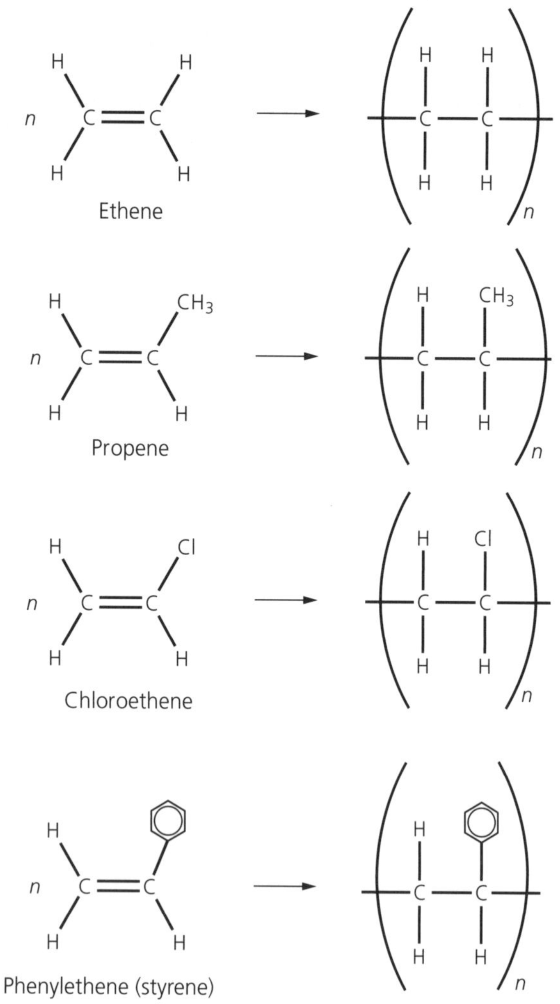

Ethene

Propene

Chloroethene

Phenylethene (styrene)

Figure 6.24

It is possible to deduce the repeat unit of an addition polymer and identify the monomer from which the polymer was produced (Figure 6.25).

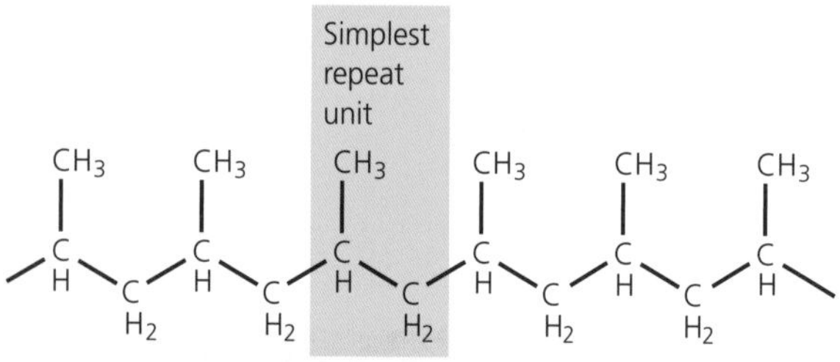

Figure 6.25

Typical mistake

When asked to draw two repeat units of the polymer formed from propene, a common incorrect response is:

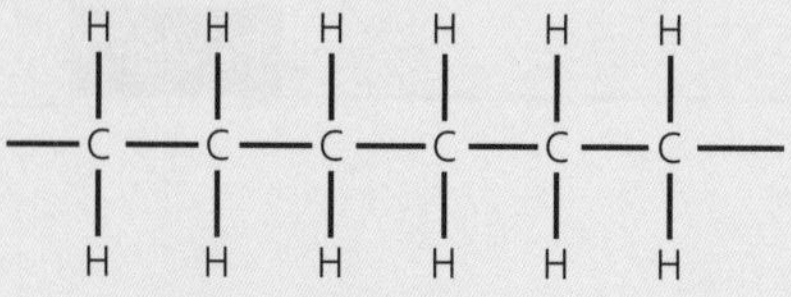

when it should be

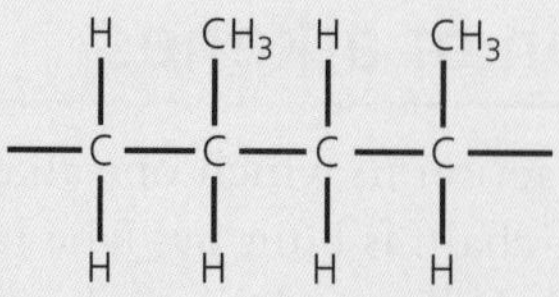

It is worth remembering that two repeat units of any addition polymer will always have a central backbone containing four carbon atoms.

Exam practice answers and quick quizzes at **www.hoddereducation.co.uk/myrevisionnotes**

Now test yourself

10 Alkenes can undergo addition polymerisation. Write an equation to show the polymerisation of but-1-ene. Draw two repeat units of the polymer.

Answer on p. 118

Polymers are an essential part of everyday life and have a wide variety of uses.

Table 6.2 Uses of polymers

Polymer	Uses
Poly(ethene)	Bags, insulation of electrical cables, bottles
Poly(propene)	Food boxes, clothing, ropes, carpets
Poly(chloroethene) — unplasticised	Water pipes, credit cards, window frames
Poly(chloroethene) — plasticised	Raincoats, shower curtains, packaging films
Poly(phenylethene)	Styrofoam cups, fast-food containers, refrigerator insulation, packaging, telephones, flowerpots

The widespread use of these polymers has created a major disposal problem. The bonds in addition polymers are strong, non-polar covalent bonds, making most polymers resistant to chemical attack. As they are not broken down by bacteria, they are often referred to as being **non-biodegradable**.

Plastic waste is usually buried in landfill sites where it remains unchanged for decades. This means that local authorities have to find more and more landfill sites.

Alternatives to using landfill sites include:
- **incineration** Polymers are hydrocarbon based and are therefore potentially good fuels. When burnt, they release useful energy. Some plastics, such as PVC, also produce toxic gases (e.g. HCl) and the incinerators have to be fitted with gas-scrubbers.
- **recycling** Polymers can be recycled and used as feedstock for the production of new polymers. Different types of polymers have to be separated because recycling a mixture of polymers would produce an inferior plastic product.

Biodegradable polymers offer a better solution. Chemists are working to try and develop new polymers that have suitable properties. The aim is to create a polymer that contains an active functional group that can be attacked by bacteria. Other options are based on condensation polymers, which you will meet if you study chemistry at A2.

Exam practice

1 (a) Ethane, C_2H_6, reacts with Cl_2 in the presence of sunlight to form a mixture of chlorinated products. One possible product is $C_2H_4Cl_2$.

    (i) State the type of mechanism involved in this reaction. [1]

    (ii) The initiation step involves the homolytic fission of the Cl–Cl bond. What is meant by the term *homolytic fission*? [1]

    (iii) Name the two possible isomers of $C_2H_4Cl_2$. [2]

  (b) When $C_2H_4Cl_2$ is treated with aqueous NaOH, it undergoes substitution reactions to form both $C_2H_4(OH)Cl$ and $C_2H_4(OH)_2$.

    (i) State, and explain, the role of the $OH^-(aq)$ in these reactions. [2]

    (ii) Draw two possible isomers of $C_2H_4(OH)_2$ [2]

2 (a) Steroids are compounds that contain four rings. Cholesterol is a steroid and it has the following percentage composition by mass: C, 83.93%; H, 11.92%; O, 4.15%.

    Show that this is consistent with the molecular formula $C_{27}H_{46}O$ and that cholesterol has a relative molecular mass of 386. [3]

  (b) Cholesterol has the structure shown in Figure 6.26. **R** represents an alkyl group.

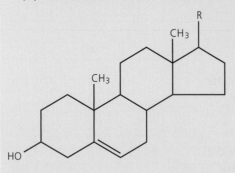

Figure 6.26

    Identify the *two* functional groups present in cholesterol. [2]

  (c) All of the compounds shown in Figure 6.27 have similar structures to cholesterol.

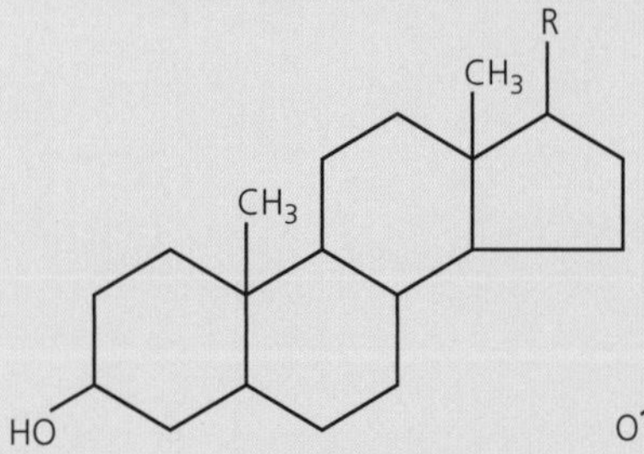

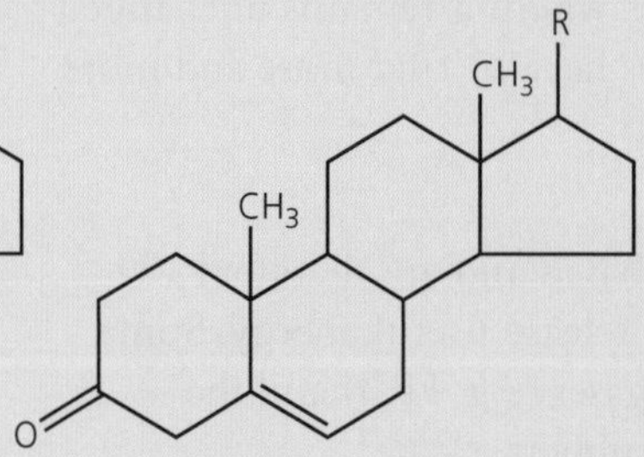

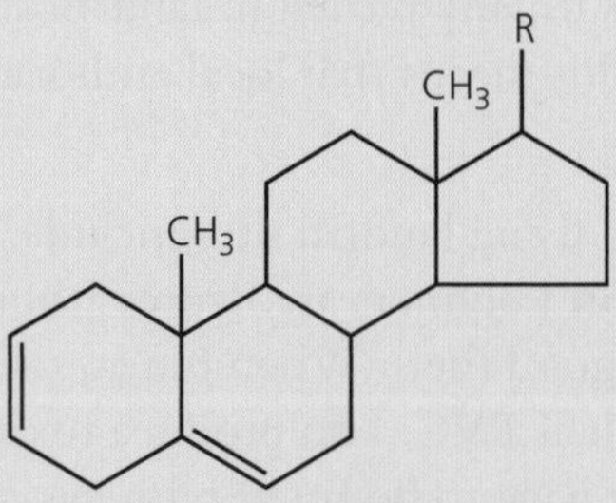

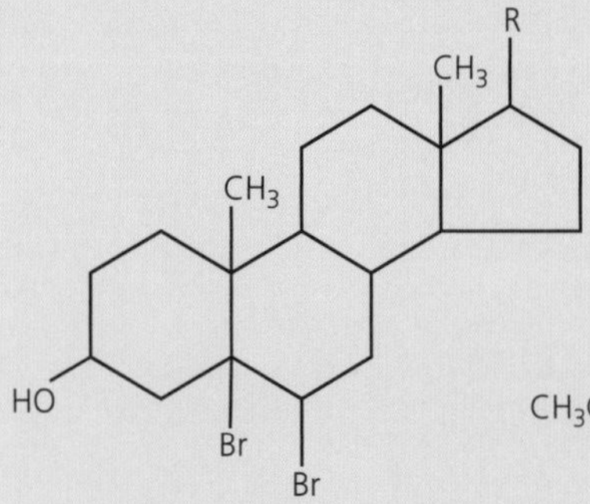

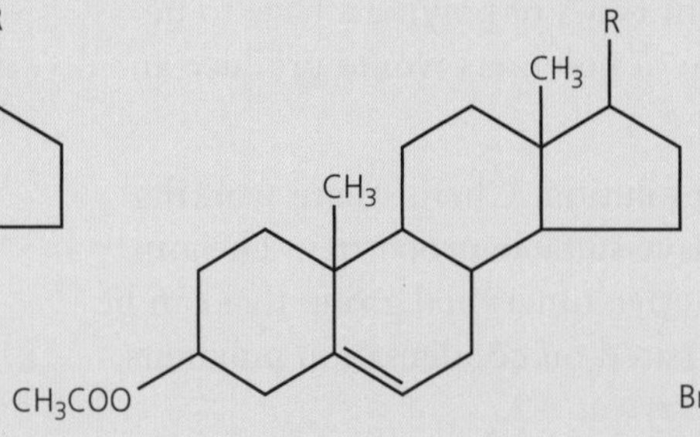

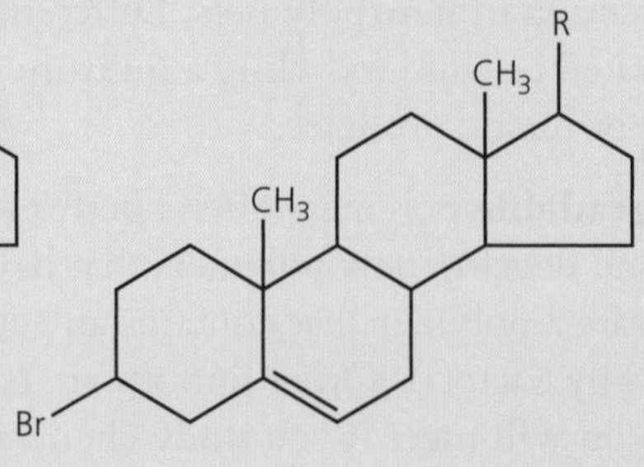

Figure 6.27

Suggest, by letter, which compound(s) might be made by reacting cholesterol with each of the following. Each letter may be used once, more than once or not at all.
(i)   bromine
(ii) ethanoic acid
(iii) hydrogen
(iv) acidified dichromate(VI) ions
(v) concentrated sulfuric acid at about 170°C   [5]

3 (a) The hydrocarbons in crude oil can be separated by fractional distillation. Explain what is meant by the terms:
   (i) hydrocarbons   [1]
   (ii) fractional distillation   [1]

(b) Dodecane, $C_{12}H_{26}$, can be isolated by fractional distillation.
   (i) Calculate the percentage composition by mass of carbon in dodecane.   [2]
   (ii) Dodecane can be cracked into octane and ethene only. Write a balanced equation for this reaction.   [1]

(c) Isomerisation of octane produces a mixture of isomers.
   (i) Name the isomers of C_8H_{18}.

A

B

C

   (ii) Isomers, **A**, **B** and **C** can be separated by fractional distillation. State the order, lowest boiling point first, in which they would distil.   [1]
   (iii) Justify the order stated in c(ii).   [2]
   (iv) Write a balanced equation for the *complete* combustion of octane, C_8H_{18}.   [2]
   (v) Why do oil companies isomerise alkanes such as octane?   [1]

4 The fractions from crude oil are processed further by cracking, reforming and isomerisation. Outline, with the aid of suitable examples and equations, each of these processes. Explain the industrial importance of each process.   [7]

Answers and quick quiz 6 online

ONLINE

Summary

You should now have an understanding of:
- the various ways in which organic formulae can be represented
- isomerism
- calculations used in organic chemistry
- key terms used in organic chemistry
- bonding, shape and boiling points of alkanes
- hydrocarbons as fuels including fractional distillation, cracking, isomerisation and reforming

- reactions of alkanes
- radical substitution mechanism
- bonding and shape of alkenes
- isomerism, including *E/Z* isomers
- addition reactions
- electrophilic addition mechanism
- Markownikoff addition
- polymerisation and the environmental aspects of disposal of waste polymers

7 Alcohols, haloalkanes and analysis

Alcohols

Classes of alcohol

Alcohols all contain the hydroxy group, −OH, and all end in '-ol'.
Alcohols can be classified as primary, secondary or tertiary.

Using 'R' to represent any other attachment, we can identify the nature of the alcohol:

Primary alcohols all contain:

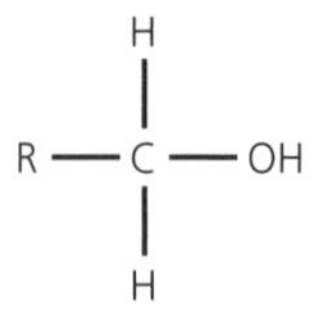

Secondary alcohols all contain:

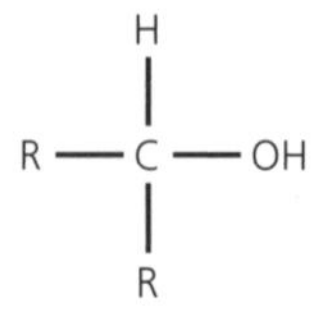

Tertiary alcohols all contain:

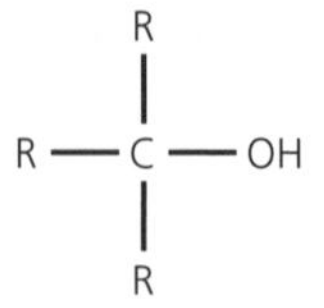

Examples of each type of alcohol are shown in Figure 7.1.

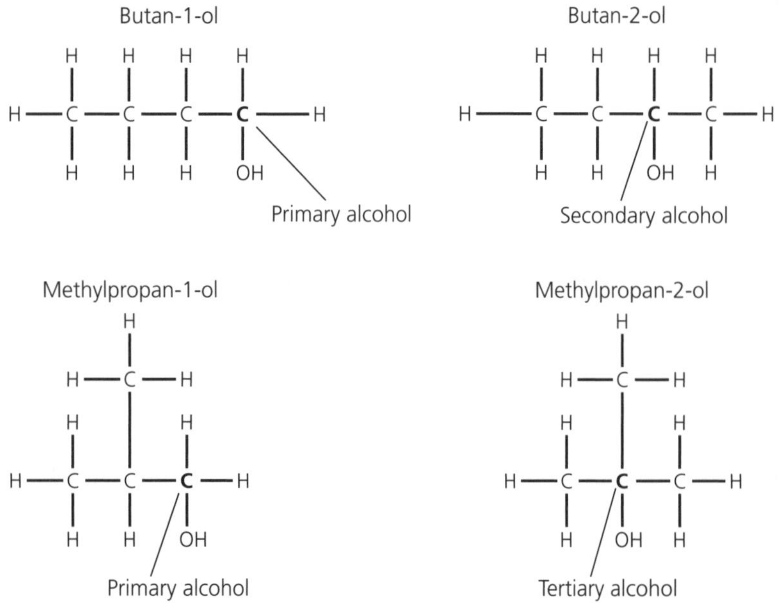

Figure 7.1

Exam practice answers and quick quizzes at **www.hoddereducation.co.uk/myrevisionnotes**

Properties of alcohols

REVISED

Alcohols have relatively high boiling points. Hydrogen bonding decreases the volatility and, therefore, results in an increase in boiling point.

Methanol and ethanol are freely miscible with water. When mixed, some of the hydrogen bonds in the individual liquids are broken, but they are then replaced by new hydrogen bonds between the alcohol and water. Miscibility with water decreases with increasing relative molecular mass of the alcohol (Figure 7.2).

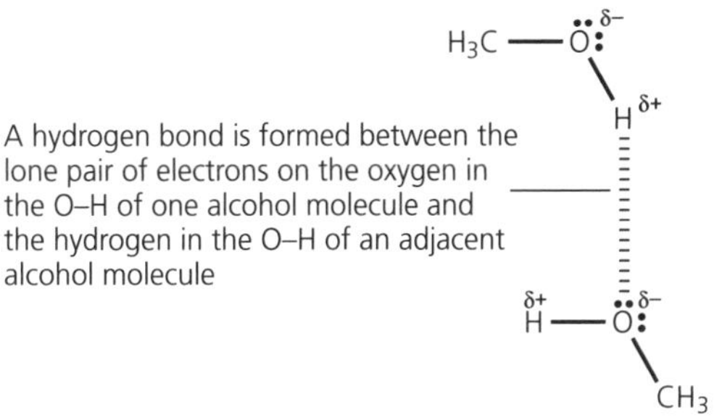

Exam tip

All alcohols can be represented by the formula R–OH, where R is the alkyl group. It follows that hydrogen bonding in any alcohol can be shown by drawing:

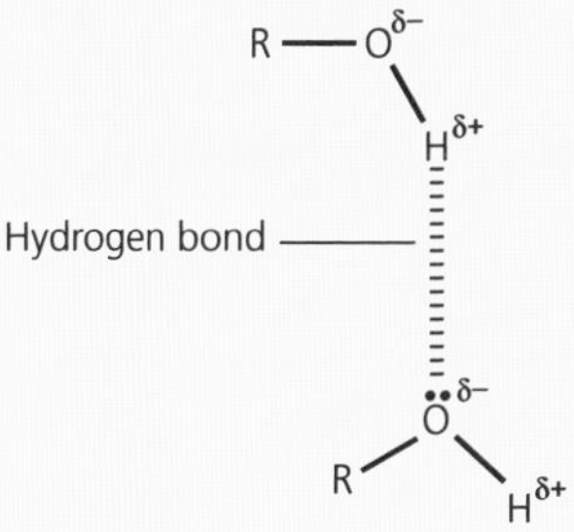

A hydrogen bond is formed between the lone pair of electrons on the oxygen in the O–H of one alcohol molecule and the hydrogen in the O–H of an adjacent alcohol molecule

Figure 7.2

Now test yourself

TESTED

1 Name each of the alcohols in Figure 7.3.

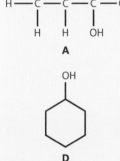

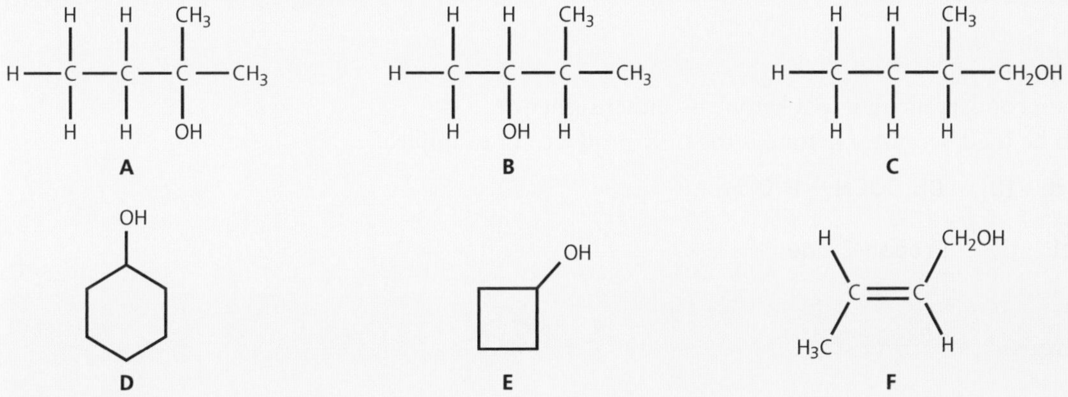

A   **B**   **C**

D   **E**   **F**

Figure 7.3

Answer on p. 118

Reactions of alcohols

REVISED

Combustion

Alcohols burn to produce carbon dioxide and water:

$$C_2H_5OH + 3O_2 \rightarrow 2CO_2 + 3H_2O$$

Oxidation

Alcohols are oxidised using the oxidising mixture $Cr_2O_7^{2-}/H^+$ (e.g. $K_2Cr_2O_7/H_2SO_4$). Oxidation reactions differ depending on the classification of the alcohol.

Each oxidation reaction is accompanied by a distinctive colour change from orange to green. Balanced equations for the oxidation reactions are

Exam tip

When writing equations for the combustion of alcohols many students forget about the O in the alcohol and write the equation for ethanol as:

$$C_2H_5OH + 3\tfrac{1}{2}O_2 \rightarrow 2CO_2 + 3H_2O$$

rather than:

$$C_2H_5OH + 3O_2 \rightarrow 2CO_2 + 3H_2O$$

written using [O] to represent the oxidising agent. Water is always formed as a co-product.

When oxidising a primary alcohol, the choice of apparatus is important. **Refluxing** produces a carboxylic acid; **distillation** produces an aldehyde.

Oxidation of a primary alcohol to an aldehyde — the more volatile aldehyde is separated out during the distillation process. An example is:

$$CH_3OH + [O] \rightarrow HCHO + H_2O$$

Methanol   Methanal

$$
\begin{array}{c}
\quad\quad O \\
\quad\quad \parallel \\
H - C \quad \text{Methanal} \\
\quad\quad \backslash \\
\quad\quad H
\end{array}
$$

Oxidation of a primary alcohol to a carboxylic acid — refluxing is used to ensure that volatile components (such as the aldehyde) do not escape and to ensure complete oxidation. An example is:

$$CH_3CH_2OH + 2[O] \rightarrow CH_3COOH + H_2O$$

Ethanol   Ethanoic acid

$$
\begin{array}{c}
\quad\quad\quad O \\
\quad\quad\quad \parallel \\
H_3C - C \quad \text{Ethanoic acid} \\
\quad\quad\quad \backslash \\
\quad\quad\quad OH
\end{array}
$$

Oxidation of a secondary alcohol to a ketone — either reflux or distillation can be used because a ketone is the only product, for example:

$$CH_3CHOHCH_3 + [O] \rightarrow CH_3COCH_3 + H_2O$$

Propan-2-ol   Propan-2-one

$$
\begin{array}{c}
\quad\quad\quad O \\
\quad\quad\quad \parallel \\
H_3C - C \quad \text{Propanone} \\
\quad\quad\quad \backslash \\
\quad\quad\quad CH_3
\end{array}
$$

Typical mistake

Oxidation of a primary alcohol to form a carboxylic acid is often incorrectly shown as:

$$CH_3CH_2OH + [O] \rightarrow CH_3COOH + H_2$$

which looks good as the symbols balance, *but* it is important to remember that water is *always* formed and the equation should be:

$$CH_3CH_2OH + 2[O] \rightarrow CH_3COOH + H_2O$$

Exam practice answers and quick quizzes at **www.hoddereducation.co.uk/myrevisionnotes**

Dehydration (or elimination)

An alcohol reacts with hot concentrated sulfuric acid or hot pumice/Al_2O_3 to form an alkene and water (Figure 7.4).

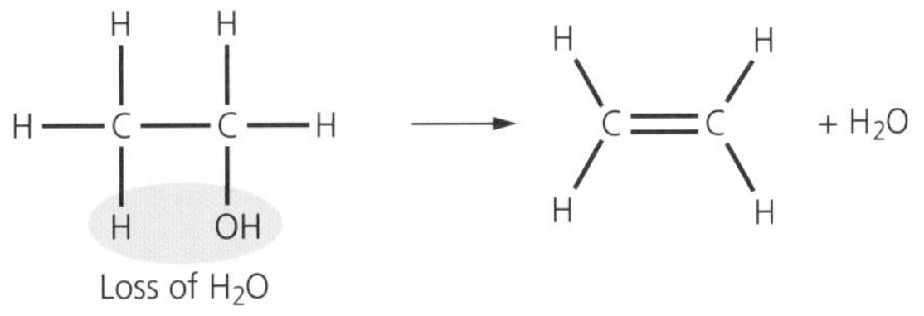

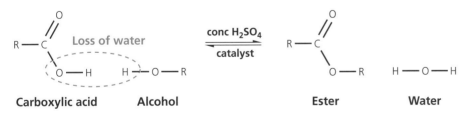

+ H2O
Loss of H2O

For alcohols like butan-2-ol it is possible to lose water in two ways:

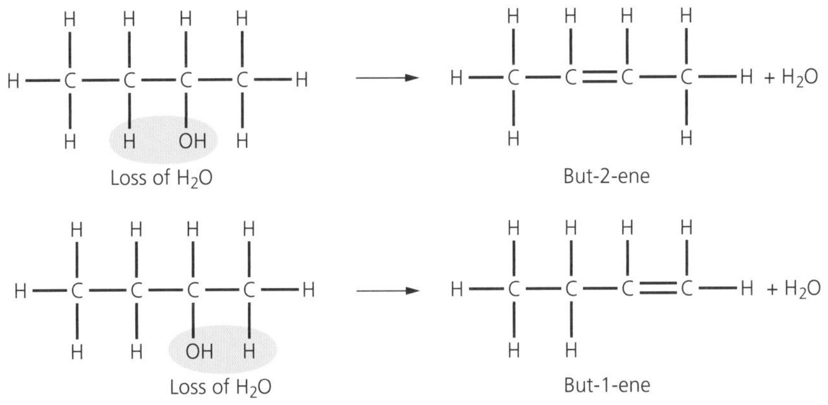

+ H2O
Loss of H2O
But-2-ene
+ H2O
Loss of H2O
But-1-ene

Figure 7.4

Substitution with halide ions

An alcohol reacts with halide ions (Cl^-, Br^- or I^-) in the presence of an acid to form haloalkanes:

$$H_3C{-}OH \xrightarrow[H_2SO_4]{NaBr} H_3C{-}Br$$

This can be represented by the equation:

$$H_3C{-}OH + Br^- \rightarrow H_3C{-}Br + H_2O$$

> **Revision activity**
>
> On three postcards use propan-1-ol, propan-2-ol and 2-methylpropan-2-ol and devise spider diagrams of the reactions of alcohols.

There is one other important reaction of alcohols. All alcohols — primary, secondary and tertiary — can react with carboxylic acids to form esters. This is covered fully in the second year of the A-level course. The equation for the formation of an ester is given in Figure 7.5.

Loss of water
conc H2SO4
catalyst
Carboxylic acid
Alcohol
Ester
Water

Figure 7.5

Now test yourself

2 Write a balanced equation for each of the following:
   (a) the complete combustion of propan-1-ol
   (b) the dehydration of pentan-3-ol
   (c) the oxidation of butan-2-ol (use [O] to represent the oxidising agent)
3 Explain what is meant by the terms *reflux* and *distillation*.
4 Explain, with the aid of equations, why the dehydration of pentan-3-ol gives one alkene but the dehydration of pentan-2-ol gives a mixture of three alkenes.

Answers on pp. 118 and 119

Haloalkanes

Classification of haloalkanes

Like alcohols, haloalkanes are subdivided into primary, secondary and tertiary. The rules for classification are the same. If the carbon atom that is bonded to the halogen (X) is bonded to one other carbon atom only, then the compound is a primary haloalkane. If the carbon atom in the C–X bond is bonded to two other carbon atoms, the compound is a secondary haloalkane. In a tertiary haloalkane, the carbon atom in the C–X bond is bonded to three other carbon atoms.

> A **haloalkane** is a compound in which one or more hydrogen atoms of an alkane is replaced by a halogen. If one hydrogen is replaced the general formula is $C_nH_{2n+1}X$ (where X = Cl, Br or I).

The carbon–halogen bond is polar, which results in the carbon atom being susceptible to attack by a **nucleophile**. A nucleophile is an electron-pair donor.

Hydrolysis of haloalkanes

When a primary haloalkane is heated under reflux with an aqueous solution of an alkali, the haloalkane is oxidised to a primary alcohol.

Reagent: NaOH or KOH

Conditions: The solvent must be water and the reaction mixture must be heated under reflux.

Equation: $CH_3CH_2Br + NaOH \rightarrow CH_3CH_2OH + NaBr$

The hydrolysis takes place by nucleophilic substitution. When describing the mechanism, it is essential to show relevant dipoles, lone pairs of electrons and curly arrows (Figure 7.6).

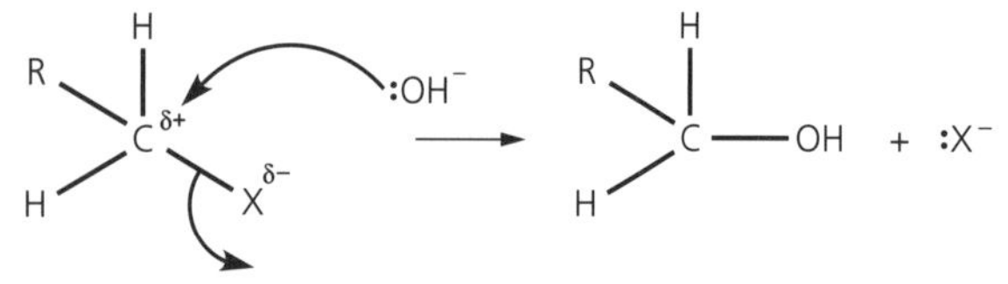

(R represents an alkyl group such as CH_3-, C_2H_5- etc.)

Figure 7.6

Rate of hydrolysis

When equal amounts of 1-chlorobutane, 1-bromobutane and 1-iodobutane are reacted under identical conditions with a hot aqueous solution containing a small amount of aqueous ethanolic silver nitrate, 1-iodobutane reacts the fastest and 1-chlorobutane reacts the slowest. The reaction can be monitored as the substituted halide ion reacts with the Ag^+ ion to produce a precipitate of white $AgCl(s)$, cream $AgBr(s)$ or yellow $AgI(s)$.

The rate of hydrolysis can be explained by comparing the carbon–halogen bond enthalpies.

Table 7.1 Bond enthalpies of carbon–halogen bonds

Bond	Bond enthalpy/kJ mol⁻¹
C–F	467
C–Cl	340
C–Br	280
C–I	240

The C—I bond is the weakest and the least energy is required to break it. The C—F is so strong that it rarely undergoes hydrolysis.

If a comparison is made of the rate of hydrolysis of primary, secondary and tertiary haloalkanes, the tertiary haloalkanes react fastest and the primary haloalkanes are the slowest. This can again be explained by the bond enthalpies: primary C—Cl bonds are the strongest and therefore the slowest, while tertiary C—Cl bonds are the weakest and therefore react the fastest.

> **Typical mistake**
>
> When asked to compare the rates of hydrolysis, many students lose marks by using incorrect language or technical terms. Students know that the iodo-compounds react fastest but often say that the difference in rate is due to the strength of the iodine bond, rather than the strength of the carbon–iodide (C–I) bond. The formula of iodine is I_2 and the strength of the I–I bond is irrelevant to this reaction.

Now test yourself

TESTED

5

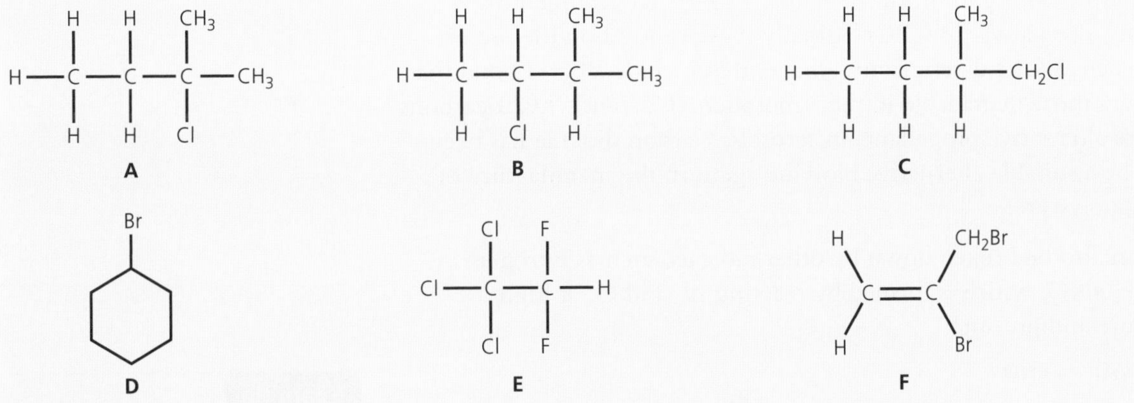

Figure 7.7

(a) Name the compounds A to F in Figure 7.7.
(b) Classify compounds A, B and C as either primary, secondary or tertiary.
(c) Write a balanced equation, and the mechanism, for the reaction of compound D with OH^-.
(d) When compound E is exposed to UV light it forms radicals. State what is meant by a radical and explain which radicals are most likely to be formed.

Answers on p. 119

Uses of haloalkanes

Haloalkanes are used in the preparation of a wide range of products including pharmaceuticals (such as ibuprofen) and polymers such as PVC (made from $CH_2=CHCl$) and PTFE (made from $F_2C=CF_2$).

CFCs

Haloalkanes were used to produce CFCs such as dichlorodifluoromethane, CCl_2F_2, and trichlorofluoromethane, CCl_3F. CFCs were developed for use in air conditioning, refrigeration units and aerosols because they are unreactive, non-flammable and non-toxic liquids of low volatility that can be readily evaporated and re-condensed. At the time of their introduction, the dangerous effect they would have on the stratosphere was not understood. It is thought that when CFCs reach the upper atmosphere they undergo photodissociation and generate chlorine radicals, $Cl\bullet$:

$$CCl_2F_2 \xrightarrow{\text{ultraviolet}} \bullet CClF_2 + Cl\bullet$$

These are extremely reactive and react with the ozone in the presence of ultraviolet light. The chlorine radical is involved in the propagation steps:

$$Cl\bullet + O_3 \rightarrow ClO\bullet + O_2$$

$$ClO\bullet + O \rightarrow Cl\bullet + O_2$$

The net reaction is:

$$O_3 + O \rightleftharpoons 2O_2$$

Chemists are working to minimise damage to the environment by researching alternatives to CFCs. Initially these centred on the use of HCFCs, such as 1,1,1,2-tetrafluoroethane, that include a C−H bond that makes them more degradable in the atmosphere. Currently hydrocarbons are used as alternative propellants in aerosols. Carbon dioxide has been found to be a suitable alternative blowing agent in the manufacture of expanded polystyrene.

Ozone can also be broken down by other radicals, such as nitrogen monoxide, $\bullet NO$, which is formed by reacting N_2 and O_2 at high temperature and pressure:

$$\tfrac{1}{2}N_2 + \tfrac{1}{2}O_2 \rightarrow \bullet NO$$

Nitrogen monoxide is found in the exhaust fumes of aircraft and can also react with the ozone by a series of reactions:

$$\bullet NO + O_3 \rightarrow \bullet NO_2 + O_2$$

$$\bullet NO_2 + O \rightarrow \bullet NO + O_2$$

The net reaction is $O_3 + O \rightleftharpoons 2O_2$ and $\bullet NO$ catalyses the breakdown of ozone.

> **Exam tip**
>
> All chemistry exams have to test 'How science works'. The way in which CFCs interact with ozone has a significant environmental impact. Make sure you know the equations.

Organic synthesis

Practical skills

Figure 7.8 details how an organic synthesis might be planned, carried out and the product analysed.

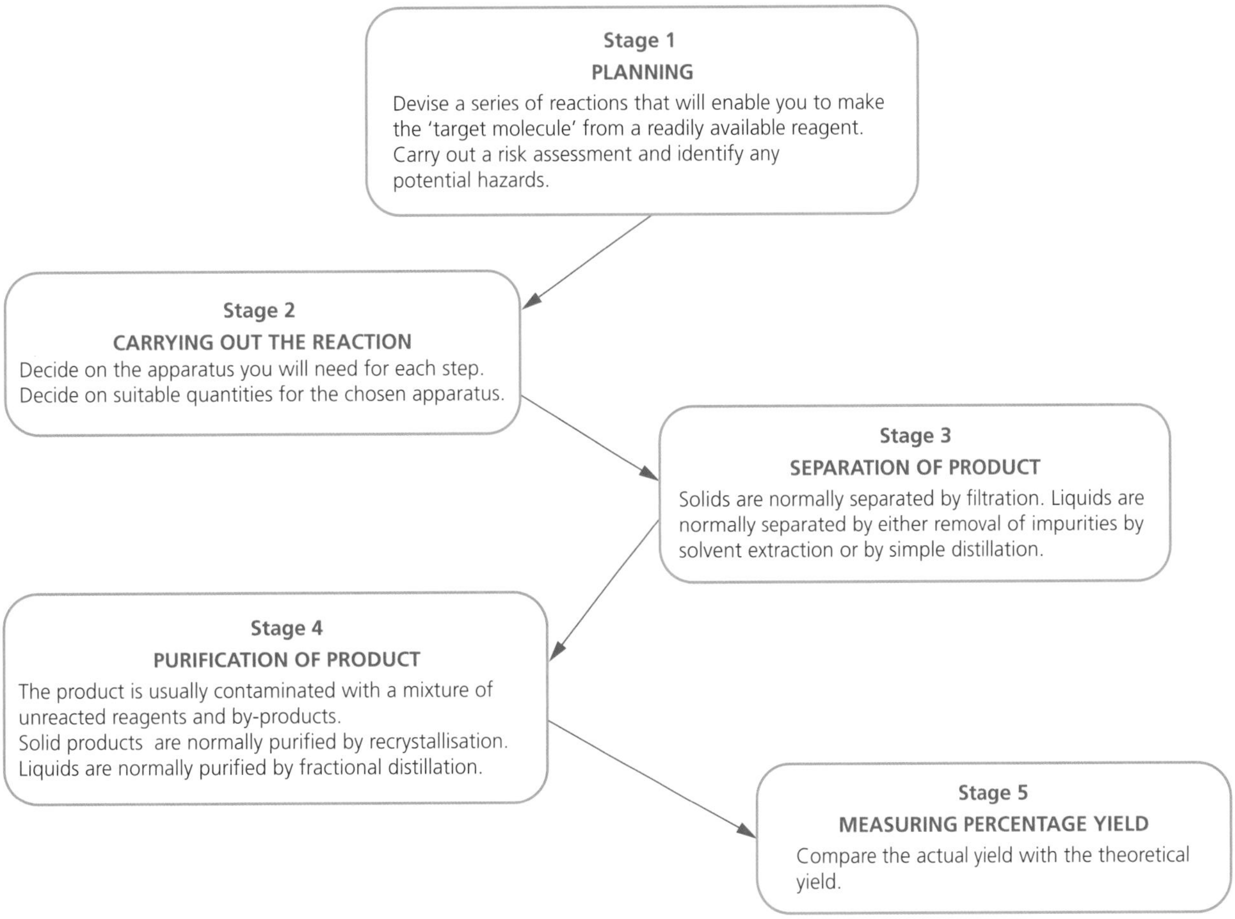

Stage 1
PLANNING
Devise a series of reactions that will enable you to make the 'target molecule' from a readily available reagent. Carry out a risk assessment and identify any potential hazards.

Stage 2
CARRYING OUT THE REACTION
Decide on the apparatus you will need for each step.
Decide on suitable quantities for the chosen apparatus.

Stage 3
SEPARATION OF PRODUCT
Solids are normally separated by filtration. Liquids are normally separated by either removal of impurities by solvent extraction or by simple distillation.

Stage 4
PURIFICATION OF PRODUCT
The product is usually contaminated with a mixture of unreacted reagents and by-products.
Solid products are normally purified by recrystallisation.
Liquids are normally purified by fractional distillation.

Stage 5
MEASURING PERCENTAGE YIELD
Compare the actual yield with the theoretical yield.

Figure 7.8 Preparation of an organic compound

Synthetic routes

Functional groups provide the key to organic molecules. Knowledge of the properties and reactions of a limited number of functional groups enables the preparation of a wide variety of organic compounds.

Table 7.2 summarises the reactions of these groups.

Table 7.2

Functional group		Type of reactions	Reagents that react
Name	**General formula**		
Alkane	C_nH_{2n+2}	Radical substitution	Cl_2, Br_2
Alkene	C_nH_{2n}	Electrophilic addition	H_2, HCl, HBr, Cl_2, Br_2 $H_2O(g)$

Functional group		Type of reactions	Reagents that react
Name	**General formula**		
Alcohol	R–OH	Oxidation	$H^+/Cr_2O_7^{2-}$
		Esterification*	RCOOH (carboxylic acids)
		Elimination	H_2SO_4
		Halogenation	$NaBr/H_2SO_4$
Haloalkane	R–Cl	Nucleophilic substitution	Common nucleophiles include: $:OH^-$, $*:NH_3$, $*:CN^-$
		Hydrolysis	

*Only likely to be tested after the second year of the A-level course

Organic chemists often start by examining the 'target molecule'. Then they work backwards through a series of steps to find suitable starting chemicals that are available and cheap enough (Table 7.3).

Table 7.3

Functional group	Reagent	Target functional group
Alkane	Halogen	Haloalkane
Alkene	Hydrogen halides	Haloalkanes
	Halogens	Di-haloalkanes
	Steam	Alcohol
	Hydrogen	Alkanes
Alcohols	Carboxylic acids*	Esters
	$H^+/Cr_2O_7^{2-}$	Aldehyde, ketone or carboxylic acid
	Hot concentrated H_2SO_4	Alkene
	NaBr in presence of H_2SO_4	Haloalkane
Haloalkane	NaOH(aq)	Alcohol
	NH_3 (ethanol)*	Amine
	Cyanide, $-C{\equiv}N$*	Nitrile

*Only likely to be tested after the second year of the A-level course

---

Example

Preparation of propanone starting from propene

In this case, the 'target molecule' is propanone and the 'starting molecule', propene, is an alkene.

Step 1 Start with the target molecule and identify the compounds that could readily be converted directly into the target — concentrate on the functional group.

Propanone is a ketone that can be made from the oxidation of a secondary alcohol, propan-2-ol.

Step 2 Look at your starting molecule, propene. What reactions of alkenes do you know?

You should now see a possible two-stage synthetic route from your starting molecule to the target molecule. In this case, the route can go via the alcohol.

$H_2C=CHCH_3$	→	$H_3CCH(OH)CH_3$	→	H_3CCOCH_3
starting molecule		intermediate molecule		target molecule

Exam practice answers and quick quizzes at **www.hoddereducation.co.uk/myrevisionnotes**

You will need to know the reagents and conditions for each step (Figure 7.9).

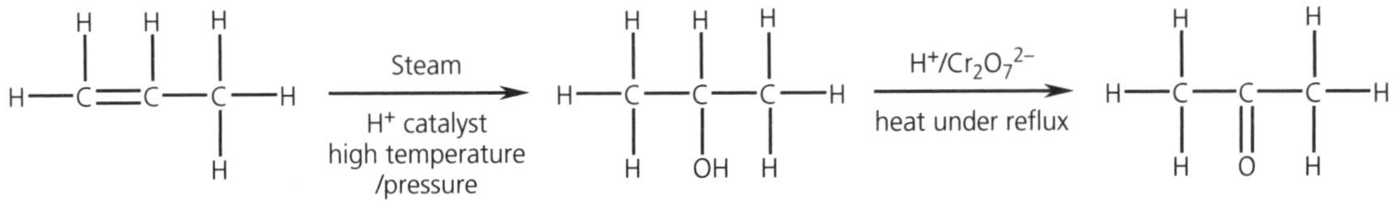

Figure 7.9

You may have to write equations for each step:

Step 1 $CH_2=CHCH_3 + H_2O \rightarrow CH_3CH(OH)CH_3$

Step 2 $CH_3CH(OH)CH_3 + [O] \rightarrow CH_3COCH_3 + H_2O$

Chemists normally seek a synthetic route that has the least number of stages and which, therefore, produces a higher yield of the product. It is rare for any one reaction to be 100% efficient; normally the percentage yield is significantly below the theoretical yield.

Now test yourself

TESTED ☐

6 Compound A decolorises bromine and when heated with acidified dichromate the dichromate turns from orange to green. Which of the three compounds is compound A most likely to be:

$CH_3CH_2CH_2OH$ $CH_3CH=CHCOOH$ $CH_3CHCHCH_2OH$

Explain your answer.
7 Devise a two-stage synthesis for converting:
   (a) methane to methanol
   (b) propene into propanone
   State the reagents and conditions needed for each conversion.

Answers on p. 119

Analytical techniques

Infrared spectroscopy

REVISED ☐

Infrared spectra can be used to identify key absorptions of alcohol, carbonyl, carboxylic acid and amine functional groups.

Absorption of infrared radiation by atmospheric gases – the greenhouse effect

The Earth is warmed mostly by the energy transmitted from the Sun. This consists largely of visible light, but there is also some ultraviolet and some infrared radiation. Most ultraviolet radiation is removed in the upper parts of the atmosphere (the stratosphere) by the ozone layer.

Over time, the surface temperature of the Earth has remained more or less constant because an equilibrium has been established between arriving and departing energy. The departing energy is almost wholly infrared radiation. Many gases in the atmosphere absorb some of this infrared energy and the absorbed infrared radiation is later released and re-emitted back to the Earth's surface.

Carbon dioxide, methane and water molecules absorb infrared radiation, as do many other gases. Any compound that contains C=O, C–H and O–H bonds will absorb infrared radiation. A wide range of gases contribute to the overall **greenhouse effect**.

The contribution of an individual gas to the greenhouse effect depends on three factors:
- its ability to absorb infrared — methane is 25 more times effective in absorbing infrared than carbon dioxide
- its atmospheric concentration — carbon dioxide is over 200 times more abundant than methane
- its residence time — CFCs can stay in the atmosphere for very many years

If the concentration of these gases is allowed to rise, the average temperature at the Earth's surface will increase.

Now test yourself

TESTED

8 State what happens to the bonds in a carbon dioxide molecule when it absorbs infrared radiation.
9 State *three* factors that influence the contribution of a gas to the greenhouse effect.

Answers on p. 120

The role of carbon dioxide as a contributor to **global warming** is the focus of much attention. This is because it is produced in huge quantities by the burning of fossil fuels.

The difficulty of providing a reliable prediction emphasises the complex nature of the chemistry of the atmosphere. Animals release carbon dioxide as they respire; plants absorb carbon dioxide during photosynthesis. Much carbon dioxide is absorbed as it dissolves into surface waters. In addition, it is not possible to predict how long an individual molecule may stay in the atmosphere — only average figures (residence time) are available. These factors make it difficult to predict the likely outcome of increasing pollution levels.

However, there is now a general acceptance of the scientific evidence explaining global warming which has prompted many governments to promote the use of renewable energy sources such as wind, solar and tidal.

Using infrared spectra to identify organic compounds

The absorptions in infrared spectra can be roughly divided into three distinct sections (Figure 7.10).

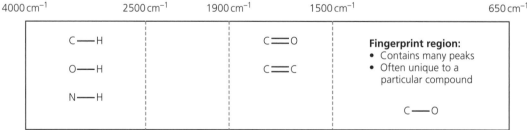

Figure 7.10

The infrared spectrum of ethanoyl chloride, CH_3COCl, shown in Figure 7.11, has an absorption due to the C–Cl bond, which occurs in the fingerprint region.

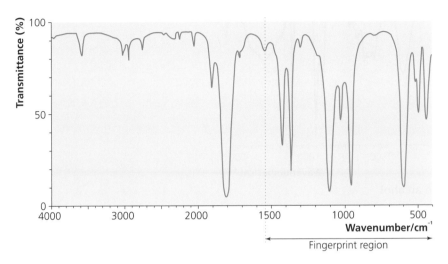

Figure 7.11

However, because there are so many peaks in the fingerprint region they are difficult to assign and it is almost impossible to tell, at a glance, which peak is due to the C–Cl bond.

The key absorptions that you may be expected to identify are given in Table 7.4.

Table 7.4 Key infrared absorptions

Bond	Location	Wavenumber range/cm^{-1}
*C–X	Haloalkanes (X = Cl, Br or I)	500–800
*C–F	Fluoroalkanes	1000–1350
*C–O	Alcohols, esters, carboxylic acids	1000–1300
C=C	Alkenes	1620–1680
C=O	Aldehydes, ketones, carboxylic acids, esters, amides	1640–1750
C–H	Any organic compound with a C–H bond	2850–3100
O–H	Carboxylic acids	2500–3300 (very broad)
O–H	Alcohols	3200–3600 (broad)
*These occur in the fingerprint regions and are difficult to assign		

You might be expected to distinguish between the infrared spectra of an alcohol and its oxidation products: aldehyde, ketone and carboxylic acid.

For an alcohol you need to identify *two* peaks (Figure 7.12):
- 1000–1300 cm^{-1} — all alcohols contain a C–O bond but they are difficult to assign as they are in the fingerprint region.
- 3200–3500 cm^{-1} — all alcohols contain a O–H bond, which gives rise to a broad peak and should not be confused with the small sharp peaks due to C–H bonds.

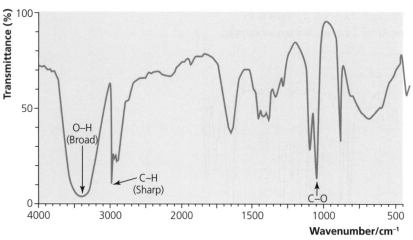

Figure 7.12 Infrared spectrum of an alcohol

For an aldehyde or a ketone you need to identify *one* peak:
- 1640–1750 cm^{-1} — all aldehydes and ketones contain a C=O bond (Figure 7.13).

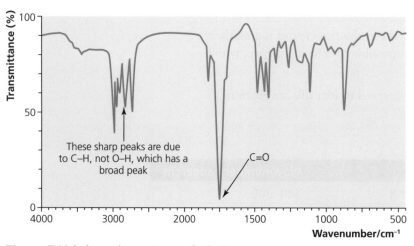

Figure 7.13 Infrared spectrum of a ketone

For a carboxylic acid you need to identify *two* (possibly three) peaks (Figure 7.14):
- 1640–1750 cm^{-1} — all carboxylic acids contain a C=O bond.
- 2500–3300 cm^{-1} — all carboxylic acids contain a O−H bond which is a very broad peak.
- 1000–1300 cm^{-1} — all carboxylic acids contain a C−O bond but they are difficult to assign as they are in the fingerprint region.

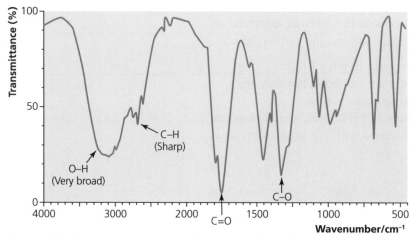

Figure 7.14 Infrared spectrum of a carboxylic acid

Infrared spectroscopy is a powerful tool in identifying a particular functional group. When identifying a specific chemical, it is usually used in conjunction with other analytical techniques such as mass spectrometry, chromatography and nuclear magnetic resonance (NMR) spectroscopy. At AS you are expected to be able to link together information obtained from an infrared spectrum and a mass spectrum.

Infrared spectrometry is also used:
● to monitor air pollution and can detect gases such as CO and NO
● in breathalysers to measure ethanol in breath

Mass spectrometry

Mass spectrometry provides evidence for the existence of isotopes. For an atom, a typical print-out from the detector looks like a 'stick-diagram'. Each stick represents an ion (isotope); the taller the 'stick' the more abundant the ion (isotope).

The mass spectrum of boron is shown in Figure 7.15.

The x-axis is labelled 'm/z', which means mass/charge. However, since the charge is 1+ it is effectively the relative mass of the ions that is recorded.

The mass spectrum of boron shows two lines, indicating there are two isotopes with mass 10 and 11.

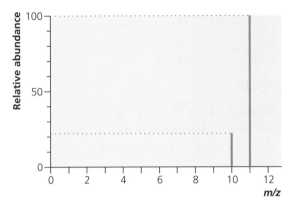

Figure 7.15 Mass spectrum of boron

The mass spectrometer is also used to analyse compounds, although the number of lines obtained may be large. This is because bombardment by electrons causes the molecule to break up and each of the fragments obtained registers on the detector. This can be an advantage as it is sometimes possible to obtain details of the molecule's structure, as well as its overall molecular mass.

The mass spectrum of propane (C_3H_8) is shown in Figure 7.16.

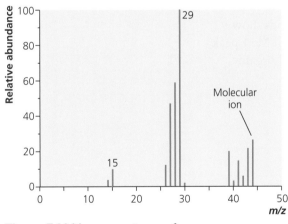

Figure 7.16 Mass spectrum of propane

Propane has relative molecular mass of 44.0 and, as expected, the peak furthest to the right of the spectrum represents the ion $C_3H_8^+$. This is called the **molecular ion peak** (or the **M** peak). However, the molecular ion is unstable and breaks down to ion fragments of the molecule, which are also detected. Some examples are labelled on the spectrum.

The peak at m/z 29 occurs because a CH_3 unit has been broken from the $CH_3CH_2CH_3$ chain and the ion $CH_3CH_2^+$ has been detected (Figure 7.17).

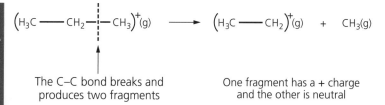

(H3C — CH2 — CH3)+(g) → (H3C — CH2)+(g) + CH3(g)
The C–C bond breaks and produces two fragments
One fragment has a + charge and the other is neutral

Figure 7.17

The peak at *m/z* 15 represents a CH_3^+ ion, and it is possible to suggest the identity of all other peaks in the spectrum (Figure 7.18).

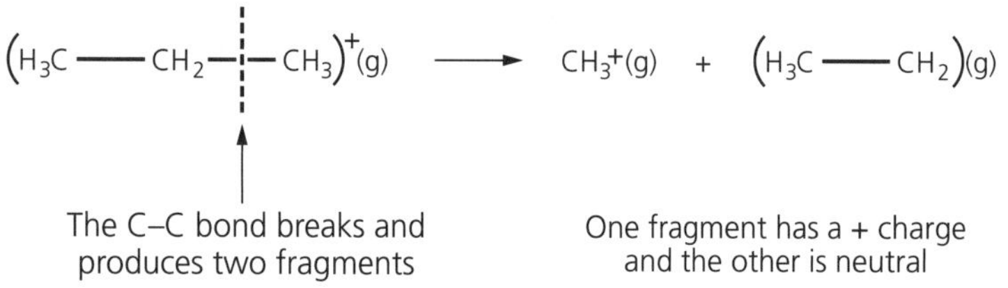

(H3C — CH2 — CH3)+(g) → CH3+(g) + (H3C — CH2)(g)
The C–C bond breaks and produces two fragments
One fragment has a + charge and the other is neutral

Figure 7.18

Fragmentation leads to a large number of peaks, giving a 'fingerprint' of the molecule. In conjunction with a computer using a spectral database, this enables a particular chemical to be identified.

For the exam you may be expected to identify a few common fragment ions, such as those in Table 7.5.

Table 7.5

m/z value	Ion responsible
15	$CH_3^+(g)$
29	$CH_3CH_2^+(g)$
43	$CH_3CH_2CH_2^+(g)$
Alkyl chains extend by CH_2 so it is possible that you will get peaks at 57, 71 etc.	
31	$CH_2OH^+(g)$ (primary alcohol)

You should be aware that the mass spectra of many compounds show not only the M peak (the molecular ion peak) but also a small M+1 peak due to the presence of the isotope ^{13}C, which is present in all organic substances (Figure 7.19).

Relative intensity
100
80
60
40
20
0
10
15
20
25
30
35
40
45
m/z

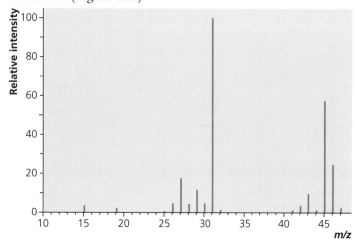

Figure 7.19

Now test yourself

10 The infrared spectrum and the mass spectrum of compound A are shown in Figures 7.20 and 7.21.

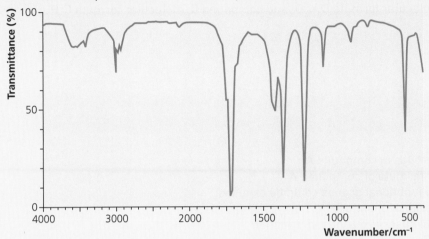

Figure 7.20 Infrared spectrum of compound A

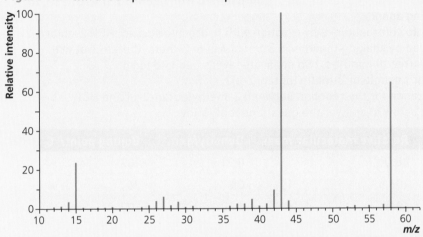

Figure 7.21 Mass spectrum of compound A

(a) Use the infrared spectrum to determine whether compound A contains:
   (i) O–H
   (ii) C=O
(b) Use the mass spectrum to determine the molar mass of compound A.
(c) Show that the molecular formula of compound A is C_3H_6O. Show *all* your working.
(d) Draw and name isomers of C_3H_6O.
(e) Use the fragmentation pattern in the mass spectrum to identify compound A. Show *all* your working.

Answers on p. 120

Revision activity

Look up the formulae of aspirin and paracetamol and suggest the *m/z* values of the molecular ion peaks of each. From the structures of aspirin and paracetamol state the range (cm⁻¹) of two absorptions that you would expect in each spectrum. Explain how you could distinguish between aspirin and paracetamol.

Exam practice

1 Compound **A** in Figure 7.22 contains two functional groups. Compound **A** can be oxidised to produce a mixture of Compound **B**, molecular formula $C_3H_4O_2$ and compound **C**, molecular formula $C_3H_4O_3$.

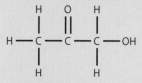

Compound **A**

Figure 7.22

  (a) Identify the functional groups in compound **A**. [3]
  (b) (i) State the molecular formula of compound **A**. [1]
     (ii) Identify which of the functional groups could be oxidised. [1]
     (iii) Suggest a suitable oxidising mixture [2]
     (iv) State what you would observe during the oxidation [1]
     (v) Identify compound **B**. [1]
     (vi) Write a balanced equation for the formation of compound **C** from compound **A**. Use [O] to represent the oxidising agent. [2]

2 Alcohols can be converted into chloroalkanes by reaction with hydrochloric acid, HCl. 2-chloro-2-methylbutane can be prepared by shaking together $5.3\,cm^3$ $(4.4\,g)$ of 2-methylbutan-2-ol with $20\,cm^3$ of concentrated HCl. After 10 minutes, two separate layers begin to form.

  (a) (i) What is the molecular formula of 2-methylbutan-2-ol? [1]
     (ii) Write a balanced equation for the reaction between 2-methylbutan-2-ol and HCl. [1]
  (b) Use the data in the table below to answer the questions that follow.

Compound	Relative molecular mass	Density/g cm⁻³	Boiling point/°C
2-methylbutan-2-ol	88.0	0.81	102
2-chloro-2-methylbutane	106.5	0.87	86
Water	18.0	1.00	100

    One of the layers is aqueous and the other contains the organic product. Suggest whether the upper or lower layer is likely to contain the organic product. Explain your reasoning. [1]
  (c) The organic layer was shaken with a dilute solution of sodium hydrogencarbonate, $NaHCO_3$. A gas was given off. Identify the gas. Suggest the chemical that could have reacted with the $NaHCO_3$ to form the gas. [2]
  (d) The resulting impure organic liquid was dried with anhydrous calcium chloride and then distilled. $3.73\,g$ of pure 2-chloro-2-methylbutane were produced.
     (i) At what temperature would you expect the *pure* organic product to distil? [1]
     (ii) Calculate the percentage yield of 2-chloro-2-methylbutane in this experiment. [3]
     (iii) Calculate the atom economy of the reaction. [2]

3 At one time, CFCs were used widely. They are now banned from production because of their damaging effect on the ozone layer.
  (a) One of the reasons CFCs used to be manufactured was for use as propellants in aerosols. Give *three* properties of CFCs that made them suitable for this purpose. [3]
  (b) CFCs damage the ozone layer by upsetting the equilibrium between oxygen and ozone in the stratosphere.
     (i) Write an equation for the equilibrium [1]
     (ii) Explain how this equilibrium is maintained in the stratosphere. [2]
     (iii) By using appropriate equations, explain how the CFC of formula CF_3Cl could deplete ozone. [3]
     (iv) Explain why a single molecule of CF_3Cl can cause the destruction of many molecules of ozone. [2]
  (c) The use of CFCs is now widely banned. However, it is expected that destruction of the ozone layer will continue for many more years. Explain why this is the case. [2]

Answers and quick quiz 7 online

ONLINE

Summary

You should now have an understanding of:
- reactions of alcohols including combustion, oxidation, elimination and esterification
- hydrolysis of haloalkanes
- nucleophilic substitution mechanism
- CFCs and ozone
- how to devise a two-stage synthesis
- absorption of infrared radiation by atmospheric gases
- how to recognise absorptions due to O–H and C=O bonds in infrared spectra
- how to determine the molar mass of a molecule by using the molecular ion peak in a mass spectrum

Now test yourself answers

Chapter 1

1 (a) extremely flammable

(b) dangerous for the environment

(c) explosive

(d) corrosive

(e) toxic

(f) oxidising

(g) irritant

(h) radioactive

2 1 with F, 2 with H, 3 with D, 4 with A, 5 with C, 6 with I, 7 with B, 8 with E, 9 with G

3 (a) (i) 2%

(ii) 1%

(iii) 0.2%

(iv) 0.48%

(v) 0.1%

(b) (i) 16.7%

(ii) 3.3%

(To measure a temperature rise the initial and final temperatures have to be measured — so the error is doubled.)

4 (a) (i) 735

(ii) 698

(iii) 0.000346

(b) (i) 7.3×10^2

(ii) 7.0×10^2

(iii) 3.5×10^{-4}

5 The student should ignore the rough reading and the $24.50\,cm^3$. The average titre is $(23.50 + 23.60)/2 = 23.55\,cm^3$, but this is usually quoted to 1 decimal place, so the average titre is $23.6\,cm^3$.

Chapter 2

1 ^{16}O has 8p, 8n and 8e

$^{23}Na^+$ has 11p, 12n and 10e

$^{19}F^-$ has 9p, 10n, 10e

2 $\dfrac{(85 \times 72.2) + (87 \times 27.8)}{100} = \dfrac{6137 + 2418.6}{100} = \dfrac{8555.6}{100}$

$= 85.556 = 85.6$ (3 s.f.)

3 $\dfrac{10x + 11(100 - x)}{100} = 10.8$

Rearrange to find $x = 1100 - 1080 = 20$

Hence 20% ^{10}B and 80% ^{11}B

4 (a) $Mg(OH)_2 = 24.3 + 16.0 + 16.0 + 1.0 + 1.0 = 58.3$

(b) $Na_2SO_4.10H_2O = 142.1 + 180.0 = 322.1$

($Na_2SO_4 = 23.0 + 23.0 + 32.1 + 64.0 = 142.1$ and $10H_2O = 10 \times 18.0 = 180.0$)

5 (a) Isotopes are atoms of the same element that have different numbers of neutrons.

(b) ^6Li has 3p, 3e and 3n; ^7Li also has 3p and 3e, but has 4n.

(c) $\dfrac{(12 \times 6) + (88 \times 7)}{100} = \dfrac{688}{100} = 6.88$

6 (a) $MgCl_2$

(b) $Al_2(SO_4)_3$

7 Rb_2SO_4

8 $MnBr_2$

9 (a) $ZnO(s) + 2HCl(aq) \rightarrow ZnCl_2(aq) + H_2O(l)$

(b) $CH_4(g) + 2O_2(g) \rightarrow CO_2(g) + 2H_2O(l)$

10 (a) Ag_2SO_4

(b) $Al(NO_3)_3$

(c) $Fe(NO_3)_3$

11 (a) $\dfrac{8.0}{32.1} = 0.25$

(b) $\dfrac{1.68}{56.1} = 0.0299$

12 (a) Use $m = n \times M$, molar mass of $AlCl_3 = 133.5$, $m = 0.04 \times 133.5 = 5.34\,g$

(b) Use $m = n \times M$, molar mass of $Al(OH)_3 = 78$, $m = 0.45 \times 78 = 35.1\,g$

13 $M = \dfrac{m}{n} = \dfrac{2.60}{0.05} = 52$; hence the element is Cr

14 $M = \dfrac{m}{n} = \dfrac{2.432}{0.02} = 121.6 = M(OH)_2$

$2 \times OH = 2 \times 17 = 34$; hence mass of X is $121.6 - 34 = 87.6 = Sr$

15 $50\,cm^3$ nitrogen and $25\,cm^3$ oxygen

16 (a) $7.33\,g$

(b) $0.625\,g$

17 $PV = nRT$, $n = \dfrac{PV}{RT} = \dfrac{101 \times 0.610}{8.314 \times 310} = 0.0239\,mol$

molar mass $= \dfrac{3.81}{0.0238} = 159.4\,g\,mol^{-1}$

Bromine, Br_2, has molar mass $= 2 \times 79.9 = 159.8\,g\,mol^{-1}$

18 (a) $0.1\,mol$

(b) 2.5×10^{-3} or $0.0025\,mol$

19 moles of NaOH = 0.05; mass of NaOH = $2.0\,g$

20 (a) Diluted by factor of 10, hence concentration $= 0.2\,mol\,dm^{-3}$

(b) Diluted by factor of 5, hence concentration = $0.4 \, mol \, dm^{-3}$

21(a) $K_2CO_3(s) + 2HCl(aq) \rightarrow 2KCl(aq) + CO_2(g) + H_2O(l)$

(b) $1.00 \times 10^{-3} \, mol$

(c) $2.00 \times 10^{-3} \, mol$

(d) $0.07 \, mol \, dm^{-3}$

22 moles of ethanol $= \dfrac{4.60}{46.0} = 0.10$

moles of ethyl methanoate $= \dfrac{5.92}{74.0} = 0.08$

% yield $= \dfrac{0.08}{0.10} \times 100 = 80.0\%$

23 molar mass of desired product $= 32.0 \, g \, mol^{-1}$

molar mass of all products $= 32.0 + 119 = 151 \, g \, mol^{-1}$

atom economy $= \dfrac{32.0}{119} \times 100 = 21.2\%$

24(a) $Ca(NO_3)_2$

(b) $Al_2(SO_4)_3$

(c) $(CH_3COO^-)_2Mg^{2+}$

25(a) full equation $\quad$ $CH_3COOH(aq) + NaOH(aq) \rightarrow CH_3COO^-Na^+(aq) + H_2O(l)$

$\quad$ ionic equation $\quad$ $H^+(aq) + OH^-(aq) \rightarrow H_2O(l)$

(b) full equation $\quad$ $CaCO_3(s) + 2HNO_3(aq) \rightarrow Ca(NO_3)_2(aq) + H_2O(l) + CO_2(g)$

$\quad$ ionic equation $\quad$ $CO_3^{2-}(aq) + 2H^+(aq) \rightarrow H_2O(l) + CO_2(g)$

26(a)

H_2O	$NaOH$	KNO_3	NH_3	N_2O
+1 −2	+1 −2 +1	+1 +5 −2	−3 +1	+1 −2

(b)

SO_4^{2-}	CO_3^{2-}	NH_4^+	MnO_4^-	$Cr_2O_7^{2-}$
+6 −2	+4 −2	−3 +1	+7 −2	+6 −2

27

Zn	+	$CuSO_4$	$\rightarrow$	Cu	+	$ZnSO_4$
0		+2 +6 −2		0		+2 +6 −2

Zn has been oxidised as its oxidation number changes from 0 to +2.

Chapter 3

1 (a) Element X is in group 15 as there is a large increase in ionisation energy between the fifth and sixth ionisations. Element X is in group 15 and in the third period so it must be phosphorus.

(b) The third ionisation energy is represented by $X^{2+}(g) \rightarrow X^{3+}(g) + e^-$

2

Factor 1	atomic radius — K is bigger than Na
Factor 2	shielding — K has more electrons than Na
Factor 3	nuclear charge — K has a greater nuclear charge than Na

Factors 1 and 2 outweigh factor 3.

3

N	$1s^2 2s^2 2p^3$
Al^{3+}	$1s^2 2s^2 2p^6$
P^{3-}	$1s^2 2s^2 2p^6 3s^2 3p^6$
Fe^{3+}	$1s^2 2s^2 2p^6 3s^2 3p^6 3d^5$

4 (a) angular

(b) pyramidal

(c) angular

(d) tetrahedral

(e) trigonal planar

(f) trigonal planar

Chapter 4

1 (a) Magnesium: atomic radius decreases across a period, shielding is the same and magnesium has the greater nuclear charge.

(b) Magnesium: atomic radius increases down a group and magnesium has less shielding. These two factors outweigh the greater nuclear charge of calcium.

(c) Neon: sodium has the larger radius and has more shielding, which outweigh the increased nuclear charge of sodium. However, neon has a full outer shell, which is extremely stable.

2 Group 4, as there is a large jump in ionisation energy between the fourth and fifth ionisations.

3 $Ca(s) + \frac{1}{2}O_2(g) \rightarrow CaO(s)$

Ca has been oxidised. Its oxidation number changes from 0 to +2.

$Ca(s) + 2H_2O(l) \rightarrow Ca(OH)_2(aq) + H_2(g)$

Ca has been oxidised. Its oxidation number changes from 0 to +2.

4 (a) $Ba(NO_3)_2$

(b) $(CH_3COO^-)_2Sr^{2+}$

(c) $Ca_3(PO_4)_2$

5 $SrCO_3(s) + 2HNO_3(aq) \rightarrow Sr(NO_3)_2(aq) + H_2O(l) + CO_2(g)$

6 $2Cl_2 + 4NaOH \rightarrow 3NaCl + NaClO_2 + 2H_2O$

7 (a) $Cl_2(g) + C_2H_2(g) \rightarrow 2C(s) + 2HCl(g)$

(b) The oxidation number of Cl in Cl_2 is 0; in HCl it is −1. This shows that the Cl has been reduced, which means that Cl_2 is the oxidising agent.

(c) The reaction should be more explosive as fluorine is a better oxidising agent than chlorine. This is because fluorine gains electrons more easily as it is smaller and there is less shielding.

8 Add each solid separately to a beaker containing HCl(aq) — the one that fizzes is $BaCO_3$.

Add the other two solids to a beaker of water — the one that is insoluble is $BaSO_4$.

Add $AgNO_3(aq)$ to the dissolved $BaCl_2(aq)$ and a white precipitate will be formed.

Chapter 5

1 Standard enthalpy change of formation, $\Delta_f H^{\ominus}$, is the enthalpy change when 1 mol of a substance is formed from its elements, in their natural state, under standard conditions of 298 K and 101 kPa.

$3C(s) + 3H_2(g) + \frac{1}{2}O_2(g) \rightarrow CH_3CH_2CHO(l)$

2 Standard enthalpy change of combustion, $\Delta_c H^{\ominus}$, is the enthalpy change when 1 mol of a substance is burnt completely, in an excess of oxygen, under standard conditions of 298 K and 101 kPa.

$CH_3COCH_3(l) + 4O_2(g) \rightarrow 3CO_2(g) + 3H_2O(l)$

3 $\Delta H = \dfrac{q}{n} = \dfrac{mc\Delta t}{n}$

$q = 75.0 \times 4.20 \times 4.6 = 1449\,J = 1.449\,kJ$

Using $n = cV$, moles of $HNO_3 = \dfrac{1.00 \times 25.0}{1000} = 0.0250$

$\Delta H = \dfrac{q}{n} = \dfrac{1.449}{0.025} = 57.96 = 58\,kJ\,mol^{-1}$

4 **Step 1** Write an equation for $\Delta_f H$ of propane:
$3C(s) + 4H_2(g) \rightarrow C_3H_8(g)$

Step 2 Construct the enthalpy triangle by writing the combustion products at the bottom:

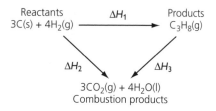

Reactants $3C(s) + 4H_2(g)$ $\xrightarrow{\Delta H_1}$ Products $C_3H_8(g)$

ΔH_2 ΔH_3

$3CO_2(g) + 4H_2O(l)$
Combustion products

Step 3 Apply Hess's law using the clockwise = anticlockwise rule

$\Delta H_1 + \Delta H_3 = \Delta H_2$, hence $\Delta H_1 = \Delta H_2 - \Delta H_3$

$\Delta H_2 = 3 \times (\Delta_c H(C(s)) + 4 \times (\Delta_c H(H_2(g))$
$= (3 \times -394) + (4 \times -286) = -2326\,kJ\,mol^{-1}$

$\Delta H_3 = -2219\,kJ\,mol^{-1}$

$\Delta H_1 = \Delta H_2 - \Delta H_3 = (-2326) - (-2219) = -107\,kJ\,mol^{-1}$

5 Activation energy is the minimum energy needed for colliding particles to react.

6 Increasing the pressure on a gaseous reaction has the same effect as increasing the concentration of the reactants because it results in a decrease in volume, so the concentration increases. Increased concentration results in an increased likelihood of a collision, so the rate of reaction increases.

7 Enthalpy profile diagram:

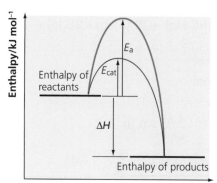

Boltzmann distribution:

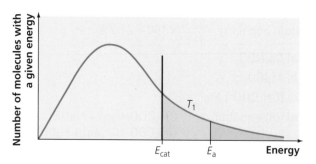

E_a is the activation energy of the uncatalysed reaction and E_{cat} is the activation energy of the catalysed reaction. A catalyst lowers the activation energy, but does not alter the ΔH value for the reaction and does not change the Boltzmann distribution. It changes the number of particles with energy greater than or equal to the new activation energy, E_{cat}. This is illustrated in the Boltzmann distribution where the areas under the curve indicate the proportion of particles with energy greater than the activation energies.

8 A heterogeneous catalyst is one that is in a different phase from the reactants.

Example: $N_2(g) + 3H_2(g) \rightleftharpoons 2NH_3(g)$ with Fe(s) as the catalyst

A homogeneous catalyst is one that is in the same phase as the reactants.

Example: esterification of ethanol and ethanoic acid (both in solution) with concentrated sulfuric acid (also in solution) as the catalyst.

Ethanoic acid   Ethanol   Ethyl ethanoate   Water

9 A *reversible* reaction is one that proceeds in both the forward and reverse directions. *Equilibrium* is achieved when the rate of the forward reaction equals the rate of the reverse reaction. Equilibrium can only be achieved in a closed system. The equilibrium is described as *dynamic* because the reactants and products react continuously; the concentrations of the reactants and the products remain constant.

10 Le Chatelier's principle states that if a closed system under equilibrium is subject to a change, the system moves to *minimise* the effect of that change.

11 (a) The temperature is increased — the ΔH value is negative, so the forward reaction is exothermic. If the temperature is increased the system will try to minimise this effect by favouring the reverse endothermic process, so the equilibrium position will move to the left.

(b) The pressure is decreased — the system will try to minimise this effect by increasing the pressure. Therefore the equilibrium position moves towards the side with the most moles of gas, i.e. to the left-hand side.

(c) $N_2O_4(g)$ is removed — the system will try to minimise this effect by increasing the amount of $N_2O_4(g)$, so the equilibrium position will move to the right.

12 $K_c = \dfrac{[N_2O_4(g)]}{[NO_2(g)]^2}$

(a) At this temperature the equilibrium lies to the left and favours the reagents.

(b) $0.0025 = \dfrac{[N_2O_4(g)]}{[1.0]^2}$; hence $[N_2O_4(g)] = 0.0025\,\text{mol dm}^{-3}$

13 $K_c = \dfrac{[NO(g)]^2[Cl_2(g)]}{[NOCl(g)]^2} = \dfrac{[(0.32)^2(0.16)]}{(3.42)^2}$

$= 1.4 \times 10^{-3}\,\text{mol dm}^{-3}$

2 3-methylpentane   2,2-dimethylpropane
  but-1-ene   propane-1,3-diol

3 2-bromopropane   but-2-ene
  2,2-dimethylpropane   3-methylpentane

4 (a) ratio of volume of alkane A:CO_2:H_2O is 100:300:400 (1:3:4)

Hence the alkane contains 3C and 8H, so the formula is C_3H_8.

(b) $10.8\,\text{g}$ H_2O is $\dfrac{10.8}{18} = 0.6\,\text{mol}$

Hence 0.1 mol of alkane B produces 0.6 mol H_2O, hence the alkane must contain 12H and therefore the formula of the alkane is C_5H_{12}.

5 Cyclobutane is C_4H_8

Equation: $C_4H_8 + Cl_2 \rightarrow C_4H_7Cl + HCl$

Mechanism:

Initiation $Cl_2 \rightarrow 2Cl\bullet$ generates radicals

Propagation $Cl\bullet + C_4H_8 \rightarrow HCl + \bullet C_4H_7$
maintains radicals

$\bullet C_4H_7 + Cl_2 \rightarrow C_4H_7Cl + Cl\bullet$

Termination $Cl\bullet + \bullet C_4H_7 \rightarrow C_4H_7Cl$ loss of radicals

Homolytic fission — breaking a covalent bond so that each atom in the covalent bond receives one of the shared pair of electrons.

Radicals are neutral atoms or groups of atoms that contain a single unpaired electron.

Substitution reactions usually occur in saturated molecules and one atom (or group of atoms) is replaced by another atom (or group of atoms).

Chapter 6

1

6 Isomers of C_6H_{14} are:

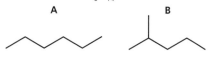

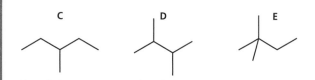

A	B
Hexane	2-methylpentane

C	D	E
3-methylpentane	2,3-dimethylbutane	2,2-dimethylbutane

D and E have the lowest boiling points.

7

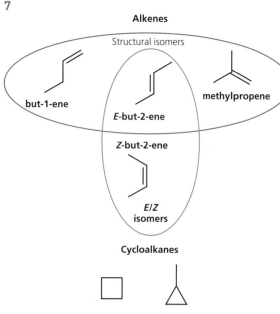

Alkenes

Structural isomers

but-1-ene

methylpropene

E-but-2-ene

Z-but-2-ene

E/Z
isomers

Cycloalkanes

cyclobutane   methylcyclopropane

8 (a) (i) $C_3H_6 + 4\frac{1}{2}O_2 \rightarrow 3CO_2 + 3H_2O$
   (ii) $CH_2CHCH_2CH_3 + H_2 \rightarrow CH_3CH_2CH_2CH_3$
   (iii) $CH_3CHCHCH_3 + H_2O \rightarrow CH_3CH_2CH(OH)CH_3$

(b) $CH_2CHCHCH_2 + 2Br_2 \rightarrow$
$BrCH_2CHBrCHBrCH_2Br$. The bromine
would be decolorised. The product is
1,2,3,4-tetrabromobutane.

(c) But-2-ene is symmetrical. Whichever way
H_2O adds across the double bond, butan-2-
ol can be the only product. But-1-ene is not
symmetrical, so it is possible to produce both
butan-1-ol and butan-2-ol.

9 (a) An electrophile is an electron-pair acceptor.

(b) $C_6H_{10} + Br_2 \rightarrow C_6H_{10}Br_2$

(c) When Br_2 reacts with an alkane the Br–Br
bond is broken homolytically and one electron
goes to each Br, producing two Br radicals.

When bromine reacts with an alkene the
Br–Br bond is broken heterolytically and the
two electrons go to one of the Br (forming a
nucleophile and an electrophile).

10

but-1-ene   poly(but-1-ene)

Two repeat units

Chapter 7

1 A = 2-methylbutan-2-ol, B = 3-methylbutan-2-
ol, C= 2-methylbutan-1-ol, D = cyclohexanol,
E = cyclobutanol, F = but-2-en-1-ol or
1-hydroxybut-2-ene

2 (a) $C_3H_7OH + 4\frac{1}{2}O_2 \rightarrow 3CO_2 + 4H_2O$

It is common to see this written incorrectly as
$C_3H_7OH + 5O_2 \rightarrow 3CO_2 + 4H_2O$ because many
students forget to count the oxygen in the alcohol.

(b) $C_5H_{11}OH \rightarrow C_5H_{10} + H_2O$

It is better to write the equation as:

(c) $C_4H_9OH + [O] \rightarrow CH_3CH_2COCH_3 + H_2O$

or

3 Reflux is a process of continuous evaporation and condensation that prevents volatile components from escaping. It does not lead to separation of products.

Distillation is a process of evaporation followed by condensation, which allows the most volatile component to be separated.

4 The dehydration of pentan-3-ol gives only one alkene, pent-2-ene, because pentan-3-ol is symmetrical. However, pent-2-ene exists as *E/Z* isomers.

Pentan-2-ol is not symmetrical and produces both pent-1-ene and pent-2-ene. Pent-2-ene exists as *E/Z* isomers.

However, pent-2-ene can exist as *E/Z* isomers

5 (a) A = 2-chloro-2-methylbutane, B = 2-chloro-3-methylbutane, C = 1-chloro-2-methylbutane, D = bromocyclohexane, E = 1,1,1-trichloro-2,2-difluoroethane, F = 2,3-dibromopropene

(b) A is tertiary, B is secondary and C is primary.

(c) $C_6H_{11}Br + OH^- \rightarrow C_6H_{11}OH + Br^-$

The marks would be for:
- correct dipoles
- curly arrow from C–Br bond to Br$^{\delta-}$
- curly arrow from :OH$^-$ to C$^{\delta+}$
- correct products

(d) A radical is a particle with an unpaired electron. In the symbol for a radical the unpaired electron is usually shown as a dot, e.g. Cl•.

The radical most likely to be formed is Cl• because the C–Cl bond is weaker than the C–F bond.

6 Compound A: decolorises Br_2, therefore is an alkene; is oxidised by acidified dichromate, therefore is an alcohol. Compound A is $CH_3CHCHCH_2OH$. (It is difficult to spot the C=C double bond: $CH_3CH=CHCH_2OH$.)

7 (a) methane → chloromethane (reagents: $Cl_2(g)$, conditions: UV light)

chloromethane → methanol (reagents: NaOH(aq), conditions: heat)

(b) propene → propan-2-ol (reagents: steam, conditions: high temperature/pressure)

propan-2-ol → propanone (reagents: acidified dichromate, conditions: heat under reflux)

8 The bonds vibrate.

9 The ability of the gas to absorb infrared radiation.

The atmospheric concentration of the gas.

The length of time the gas stays in the atmosphere — the residence time.

10 (a) (i) No broad peak at about $3000\,cm^{-1}$, therefore O–H is not present.

(ii) Peak at about $1700\,cm^{-1}$, therefore C=O (carbonyl) is present.

(b) molar mass = 58.0

(c) Compound A contains C=O, which has mass 28.0 (12.0 + 16.0).

The molar mass = 58.0, so the rest of the molecule must have a mass of 30.0.

Hence, there must be two carbon atoms (24.0) and six hydrogen atoms.

The formula is C_2H_6CO or C_3H_6O.

(d) CH_3CH_2CHO — propanal; CH_3COCH_3 — propanone

(e) Propanal forms a fragment ion, $CH_3CH_2^+$, at $m/z = 29$ and propanone does not have a fragment peak at $m/z = 29$. The fragmentation pattern confirms that compound A is propanone.

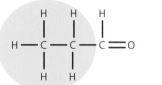

Gives fragment $CH_3CH_2^+$ which has $m/z = 29$

Does not give fragment with $m/z = 29$